The Story of Georgia's Boundaries

A Meeting of History and Geography

William J. Morton, M.D., J.D.

Georgia History Press
Atlanta, Georgia

THE STORY OF GEORGIA'S BOUNDARIES: A MEETING OF HISTORY AND GEOGRAPHY
©2009 William J. Morton

All rights reserved. No part of this book may be reproduced or transmitted in any form or by any means, electronic or mechanical, including photocopying, recording, or by any information storage and retrieval system, without written permission from the publisher. Although the information contained in this book is believed to be true and correct, the publisher and author assume no liability for errors or omissions.

Printed in the United States of America • First Edition

ISBN (hard cover): 978-0-9841596-1-1
ISBN (soft cover): 978-0-9841596-0-4
ISBN (e-book): 978-0-9841596-2-8

Library of Congress Control Number: 2009909130

Book design by Katherine Siegel • www.kartouchedesign.com
Maps by John C. Nelson • www.johnnelson.com

Georgia History Press
P.O. Box 190632 • Atlanta, Georgia 31119

The Story of
GEORGIA'S BOUNDARIES

Also by William J. Morton

Medical Malpractice: Handling Urology Cases

**To My Wife and Best Friend,
Monna Jean**

*Who very much encouraged me to write this book and listened to me
as I spent hours talking to her about events and personalities.
She gave me constructive criticism with her advice and
taught me patience and humility — both of which I sorely lack.*

Table of Contents

List of Maps & Illustrations xi

Introduction 1

I: Setting the Stage 3

    The early explorers, Spain and Portugal

    France

    England

    Virginia

    The Plymouth Company

    Jamestown

    Carolina

    The Intercolonial Wars

    Native Americans

Important Dates I 16

II: Georgia, the Thirteenth Colony 17

    The five early attempts

    The Trustees of Georgia

    The new colony

    Relations with the Creek

    First settlements

    Dissent and the Malcontents

    The War of Jenkins' Ear and Oglethorpe

The Intercolonial Wars continue

The Royal Period of Georgia

Important Dates II 36

III: The Birth of a State and a Nation. 37

The Proclamation Line and the Florida colonies

Unrest in the colonies

The path to independence

Georgia's signers

Georgia's constitution

The fight for independence

A new nation emerges

Western expansion

The Beaufort Convention

Establishing a government

The birth date of Georgia

Important Dates III 65

IV: Defining Georgia 67

Surveying

Latitude and longitude

The Pinckney Treaty

Andrew Ellicott

The Yazoo Land Fraud

Tennessee

Mississippi

Georgia land cession

The "nonexistent land"

Alabama, Tennessee, and Montgomery's Corner
James Camak
Florida
The Trail of Tears

Important Dates IV 91

V: Georgia's Boundary Disputes 93
The legal boundaries of Georgia
Resolving disputes between states
Georgia and Florida boundary disputes
Georgia and Alabama boundary disputes
Georgia and Tennessee boundary disputes
Georgia and North Carolina boundary disputes
Georgia and South Carolina boundary disputes
Conclusion

Important Dates V 149

Epilogue 151

About the Author 155

Acknowledgements 157

Bibliography 161

Index 175

List of Maps & Illustrations

After Treaty of Tordesillas, 1494. 4
Trustees' charter boundaries, 1732 21
Colony of Georgia seal 23
Savannah, 1741 26
After the Treaty of Paris, 1763. 32
Proclamation Line of 1763 38
Colony of West Florida, 1763 39
Colony of East Florida, 1763 40
Result of Gov. Wright's request to Board of Trade, 1764. 41
After the expansion of West Florida, 1767 42
The Mitchell Map. 56
After the Revolutionary War, 1783 58
After the Beaufort Convention, 1787 61
Example of latitude and longitude 69
West Florida comparison 71
After Georgia Constitution amended, 1798. 78
Northern and western boundaries
    after Articles of Agreement and Cession, 1802. 79
The disputed "nonexistent land," 1787–1811. 81
Placement of Ellicott's Rock at 35th parallel, 1811. 82
Northern and western boundaries
    with Montgomery's Corner, 1826 87
The boundaries of Georgia, 2009 95
Orr–Whitner Line, 1861 110
After U.S. Supreme Court decision, 1990 147

Introduction

The history of the New World has been examined very thoroughly over the years, but the story of Georgia's boundaries — or any state's boundaries, for that matter — receives very little attention outside of specialized arenas. A state's boundaries usually do not affect individual, ordinary people so we pay them little heed other than perhaps noticing a road sign, but consider how each of us feels about the property lines which define the space of our personal ownership and control. A state is no different and its boundaries do have real-world consequences for its citizens. The latest lawsuit between Georgia and South Carolina, only 20 years ago, underscores the importance that states give to their sovereignty, and the 2008 drought in Georgia resurrected the issue of a faulty survey of the northern boundary performed 200 years earlier which could have significantly changed the rights to water in the Tennessee River.

At first blush, a book about the boundaries may appear to be a mundane topic, but the events that influenced them and the people who measured them provide a marvelous dimension to the story. From 1732 when the charter of King George II to the Trustees of Georgia provided the first description of the boundaries of the new colony until 1838 when the last of the Cherokee were forcefully removed from their land, Georgia's boundaries changed more than a dozen times. Wars and treaties, political maneuvering and litigation, even simple human error have all affected the boundaries of the present-day state of Georgia.

Add to this drama the human aspect of the men who spent months and years walking through the woods and swamps of uncharted territory and marked the line using primitive instruments along with their knowledge of mathematics and astronomy. They endured primitive conditions and risked their lives to perform their task.

Clearly, the state's recognizable shape did not emerge intact from a vacuum. The boundaries of Georgia are truly a meeting of history and geography.

I
Setting the Stage

The story of Georgia's boundaries begins on another continent, centuries before the trustees of Georgia received the charter for their colony from King George II in 1732. European nations lusted for more power, wealth and land — their imperialism evidenced by the constant wars amongst them — and for almost 250 years after the first European set foot on the New World, the English, French, Spanish, Italians, Portuguese and Dutch made heavy investments to explore the continent. Understanding their interrelationships is the preamble to appreciating the history that literally shaped the state of Georgia.

Europe's early colonization of America was driven by ambitions to find the direct route to India, China and Japan for fabled treasures; spices such as nutmeg, cinnamon and garlic were necessary for Europeans to preserve their food, and the potential gold market would be a boon to investors. The overland route was thousands of miles long, over mountains and deserts, and subject to bandits and unfriendly nations. Marco Polo, who had made the trip 250 years earlier and lived in China for years, wrote two books about his experiences which inspired Europeans to find a shorter and safer route.

The early explorers, Spain and Portugal

Christopher Columbus, an experienced sailor and navigator, successfully persuaded Spain's Queen Isabella to sponsor his expedition looking for the direct ocean route to India. Using his knowledge of astronomy, trade winds and currents, he landed on a Bahamian island two-and-one-half months after departing Spain. Columbus spent almost three months in the New World exploring the southern archipelago of the Bahamas and then sailing further south to a huge island which he named La Española, now called Hispaniola. Because he was convinced that he had found the off-shore islands of India, he called the islands the West Indies and the local natives Indians.

Almost immediately after Columbus' triumphal discovery of the New World in 1492, bitter enemies Spain and Portugal (both Catholic countries) agreed to split up the new land which he had discovered. Although no European, other than Columbus and his crew, had ever seen or even imagined how much land was involved, both Spain and Portugal agreed to ask the Pope for his blessing and approval of a territorial division, as was the custom of the time. The 1494 Treaty of Tordesillas was a lopsided agreement which was supposed to deter any other country from gaining access to the new land. It provided that a Line of Demarcation be drawn from the top to the bottom of the Earth such that Spain and Portugal would have proper claim to the new lands. Since no one had any idea of the size

After Treaty of Tordesillas, 1494

or configuration of the earth or even where this imaginary line was, the whole idea was absurd. However, both countries agreed that undiscovered lands to the west of the line were to be Spanish possessions and those to the east of the line belonged to Portugal.

In 1500, some six years after the treaty was signed, the Portuguese explorer Pedro Cabral was blown off course while attempting to reach India by sailing around the southern tip of Africa. He landed on what turned out to be the South American continent. Believing that he was east of the Line of Demarcation, Cabral claimed the land in the name of Portugal. Twenty-five years later, Portugal returned to the area to establish a colony which they called Brazil, after a type of wood found there. Portugal has maintained a presence in Brazil ever since, which is why Portuguese is spoken in that country but nowhere else in South America.

England and France completely ignored the Treaty of Tordesillas and began to explore and establish their own colonies in the New World. England's King Henry VII granted John Cabot and his three sons financial support for a voyage to make land claims on behalf of England. In June 1497 after a six week voyage, Cabot landed near Nova Scotia or Newfoundland and claimed the land for England, although he made no contact with any natives. Cabot stayed very briefly and then returned to England to report to the king and to obtain more supplies for a second trip. Within a few months, he made a second voyage, but he and his crew were never heard from again. It would be almost a hundred years before England would resume further exploration across the ocean.

Columbus made three more voyages exploring the Caribbean islands and Central America. He took gold and precious stones from the inhabitants to give to Queen Isabella, but he never found the jewels that he thought would be so plentiful. He was very cruel to the natives and kidnapped several of them to bring back to Spain. On his last voyage, he treated his own crew so badly that they mutinied and held him captive for several weeks. Up to his death in 1506, he believed he had discovered the direct route to the riches of Asia, never appreciating the huge North American land mass that stood in his way.

The Spanish explorer Juan Ponce de León had been with Columbus on his second voyage to Hispaniola, but rather than returning to Spain, Ponce de León stayed in the Caribbean to explore and conquer the island of Puerto Rico. He traveled extensively throughout the other Caribbean islands looking for gold and minerals and was also a cruel dictator to the native Caribe inhabitants. Ponce de León eventually sailed north from the Caribbean with 200 men and three ships. On April 2, 1513, he sighted land which he called La Florida, a name loosely translated either as the "Feast of Flowers" or the "Easter Feast." After traveling south around the peninsula and up the west coast, and without coming ashore, he returned to Spain. Eight years later, he again sailed to La Florida to start a settlement with several hundred men, livestock and supplies. His exact site of landing is unknown, but it is well documented that he was wounded during an assault by angry natives and taken to Cuba, where he died. Ponce de León is buried in the Cathedral of San Juan Bautista in San Juan, Puerto Rico.

The next Spanish explorer to come to the New World was Hernando de Soto, who landed on the western side of Florida near modern-day Tampa in 1539. De Soto had already achieved significant wealth and fame from travels to Peru, where he and Francisco Pizarro had conquered the Incas, and he petitioned King Charles V of Spain for permission to conquer the Caribbean. Leaving Spain with several ships, 300 men and 200 horses, he sailed to Cuba. After a brief naval and land battle with the French, he overcame them and forced them to leave Cuba, which he then claimed for his patron.

De Soto took with him two persons whose only role was to chronicle his voyages, and it is their diaries which provide the information about his exploration in the New World. Over the next four years, de Soto's travels took him into what would become the southeastern states of Florida, Georgia and South Carolina, then westward where he discovered the Mississippi River. He proved to be a vicious and untrustworthy leader in his dealings with the Native Americans, and by the end of his time spent in the New World, half of his men were either killed from fighting the native populations or dead from various diseases. De Soto became ill during the journey and, after he died, his body was buried in the Mississippi River so that it would not be desecrated by the local inhabitants. Afterward, his crew sailed down the Mississippi into the Gulf of Mexico, landing in Mexico in 1543.

Although de Soto found none of the treasures believed to be in the New World, his expedition was the beginning of the strong Spanish presence on the north and south continents for the next 250 years. His chroniclers provided valuable information about the native populations, rivers, trails and villages which they came upon, but de Soto also set the stage for the inhumane treatment of the indigenous peoples whom the Spanish met in the New World.

The first European settlement in the New World was St. Augustine, founded by Spanish Admiral Pedro Menéndez de Avilés in 1565. For the next 100 years, Spain, with its strong Catholic religious beliefs, founded numerous missions from La Florida to the Chesapeake Bay to convert the Native Americans to Catholicism. Early 16th-century maps show Spanish La Florida as all the land west of the Allegheny Mountains and from the Gulf of Mexico to Canada.

France

France was also interested in finding the fabled Northwest Passage to India, and in 1534 Jacques Cartier made the first transoceanic voyage on its behalf. He landed on what he believed to be an offshore island of Japan or China but was, in fact, Newfoundland, close to where England's John Cabot had already explored some 40 years earlier. Cartier sailed further northward, mapping the coastal areas and exploring the islands and gulfs of Newfoundland, Labrador and Nova Scotia. He claimed all he landed on in the name of his sovereign, Francis I, not knowing — or perhaps not caring — that Cabot had already been there claiming the same lands for England. Cartier named these lands Canada, an Iroquois word meaning "village" or "settlement."

Cartier made three separate trips to the Canadian coast and up the Saint Lawrence waterway. His first trip was perhaps the most productive in that he formed excellent relationships with the local inhabitants and even brought back two teenagers with him to France.

He exhibited the youths in front of King Francis I and the French people, hoping this would persuade the Crown to continue supporting his explorations. Apparently his strategy was successful, and after receiving additional funding, Cartier returned to Canada. This time, he traveled further up the Saint Lawrence River. Upon reaching a series of rapids which he was unable to navigate, he turned around and returned to France. The Lachine ("The China") Rapids, west of Montreal, were so named because Cartier thought he had found the Northwest Passage to China.

In 1541, with an eye toward gold and diamonds rather than the route to Asia, Francis I again commissioned Cartier to return to Canada. Unlike the two previous voyages when Cartier and his men explored the region and lived on their boats, this time they built the first documented French presence on the new continent: a fort called Charlesbourg-Royal. The French were not interested in conquering the native population and only wanted to learn how to mine the area's precious metals. Unfortunately, the gold turned out to be iron pyrite, or fool's gold, and the diamonds turned out to be quartz, giving rise to the French proverb, "As false as Canadian diamonds." Over the next year, however, their relationship with the Iroquois fell apart, and Cartier abandoned the fort and returned to France where he spent the rest of his life until his death in 1557 at age 66.

Because of Cartier's journeys, France claimed a huge area of North America, which was called New France. French citizens traveled to Newfoundland and Nova Scotia, westward on the Saint Lawrence River to the Great Lakes, and down the Mississippi River to New Orleans. In 1608, Samuel de Champlain built a settlement on the Saint Lawrence River. Called Quebec City, this was the administrative center of New France. The population of New France grew to over 75,000 by the 1750s.

England

Long after the French and Spanish, almost a hundred years after Columbus, England awoke from its exploration slumber spurred by a need to intensify the search for new economic opportunities from fishing, furs and agricultural products. Although Cabot had made the initial tentative steps, it took English investors and the interest of Richard Hakluyt to recognize that exploration of the New World was important even if there was no direct route to Asia from Europe.

Hakluyt was passionate about his country's need to take a commanding global position. Educated at Eton and ordained a chaplain in 1577, he is known for devoting his life to record all written works related to exploration of the New World. Hakluyt published numerous manuscripts of French, Portuguese and Spanish explorers with the idea that he could promote English colonization of the New World. His first treatise, *Divers Discovery Touching America and the Islands Adjacent*, brought him national fame. He became an investor in joint ventures to settle new English colonies in North America and to find the Northwest passage to Asia. He also was an advisor to the successful mercantile East India Trading Company. The Scottish historian William Robertson wrote of Hakluyt, "England is more indebted for its American possessions to him than to any other man of that age."

Since the mid-16th century, Portuguese, French and Spanish ships had crossed the Atlantic to fish the rich waters off the Grand Banks. Now England was interested in not only controlling the market, but also in forcing the other countries out of the New World. In 1583, persuaded by Hakluyt, Queen Elizabeth I agreed to sponsor Sir Humphrey Gilbert's expedition to the New World. Gilbert, a member of Parliament, had a passion to find the Northwest passage to Asia and to increase England's colonization of the lands that Cabot had first discovered. He landed on St. John's Island, Newfoundland, and then explored several hundred miles to the south, claiming all the land for England. (This manner of "claiming" the land continued until the early 18th century when occupancy and military strength decided who was the true owner.) During his return to England, Gilbert and his ship were lost at sea, but documents from several other vessels on the voyage authenticate his journey.

The Gilbert family went on to play a large role in England's early exploration: Humphrey's son John was sponsored by Queen Elizabeth I for additional northwestern exploration and his other son

Raleigh was a prominent investor in the Virginia Company of London.

Virginia

Sir Walter Raleigh, the half-brother of Sir Gilbert, received a grant from Elizabeth I in 1584 to establish a presence in the New World to provide a base for English privateers to attack Spanish vessels and capture their cargo. The new colony was named Virginia, in honor of the "Virgin Queen." Virginia was a proprietary colony; that is, Raleigh and his investors, including Hakluyt, were granted the property as an investment and were to pay the Crown a percentage of any profits made from colonial commerce. This was to be the recurrent theme of England's method of gaining a foothold in the New World, and only after the proprietary colonies failed did England provide the resources for their success.

In 1585, Raleigh sent 100 people, who landed on Roanoke Island in North Carolina, to build the first English settlement in the New World. Within a year they abandoned the colony because of dwindling supplies and hostilities with the local population. Not to be deterred, Raleigh sent another, better-organized group with more supplies to the same area in 1587. The Native Americans again proved to be a serious problem; within a few weeks, at least 15 of the new arrivals were murdered. The determined settlers built a fort and dispatched the proprietary governor, John White, back to England for more supplies and militia to resist the indigenous people.

Unfortunately it took over two years for White to return to the colony, at which time he found the fort and houses intact but abandoned. The only clues were the word "Croatoan" carved into a post of the fort and "Cro" carved into a nearby tree. White attempted to determine what happened to his family and the other settlers to no avail, nor have the past 400 years provided a satisfactory answer. The Roanoke colony is now part of America's folklore. Schoolchildren learn that Virginia Dare, Governor White's granddaughter, was the first European born in the New World, and "The Lost Colony," touted as America's longest-running outdoor drama, is performed every summer on Roanoke Island.

The Plymouth Company

Twenty years after the failure of the Virginia colony, new investors received permission from England's King James I to form two separate companies with the goal of establishing settlements in that same region. Investors in the Virginia Company were given land between the 34th parallel north and the 41st parallel north (roughly from modern day Wilmington, N.C., to New York City), and investors in the Plymouth Company were granted land from the 38th parallel north to the 45th parallel north (from Baltimore to the Canadian border). The overlapping territory was created with the proviso that the two companies were not permitted to establish colonies within 100 miles of each other.

In August 1607 the Plymouth Company sent two ships with a hundred settlers from Plymouth, England, to the coast of Maine where they built the Popham Colony (named after John Popham, the Lord Chief Justice of England). This colony failed because of the harsh Maine winter weather and difficulties with the Native Americans.

Another Plymouth Company colony was established in 1620 by a group of settlers looking for religious freedom from the Church of England. These self-called Pilgrims landed at the tip of Cape Cod and named their colony Plymouth after the company which sponsored them. Despite severe hardships, illnesses and the ever-present angry natives, the colony survived for the next 70 years until it was absorbed by the Massachusetts Bay Colony in 1691.

Jamestown

The Virginia Company sponsored a well-organized group of men and ships who sailed up the James River in 1607 to settle Jamestown, destined to be the first successful English colony in the New World. The early years were difficult for these settlers, facing the same struggles as the previous colonists. After the first governor was murdered, Lord De La Warr was sent to Virginia in 1610. His control of the Native Americans, based on his strong military background, persuaded the original settlers not to give up. He stayed in Jamestown for only two years before returning to England to write a book about Virginia, leaving his deputy Sir Samuel Argall in charge. De La Warr, who remained the governor, died in 1618 on the voyage back to Jamestown to investigate complaints about Argall's harsh rule. The state of Delaware as well as the Delaware River and Bay bear his name.

Colonist John Rolfe's entrepreneurial talents were responsible for the economic survival of Jamestown. While England was spending huge sums of money buying tobacco from Spain and other European countries, Rolfe brought seed from Trinidad and successfully cultivated tobacco leaves, giving the colony a commercially viable enterprise. In addition, Rolfe's marriage to young Pocahontas, daughter of the local tribal chief, provided the settlers a period of peace with the natives.

As the Virginia Company investors began realizing profits, they put more money and other resources into Jamestown. Perhaps the most important step taken by these investors was to direct the colonists to "establish one equal and uniform government over all Virginia." The colonists formed the House of Burgesses, a legislative body modeled after England's Parliament which met once a year. Jamestown would serve as a model for the representative government on which the United States of America would be founded. Such historical luminaries as George Washington, Thomas Jefferson and Patrick Henry were in the House of Burgesses and, because of their collective experiences, played leading roles in the development of the new nation. However, because of increasing mismanagement within the Virginia Company, King James I revoked its charter, and Virginia became a crown colony in 1624. While the House of Burgesses continued to meet for the next 150 years, the original form of a representative type of government gradually became diluted as English restrictions grew.

Carolina

With the success of the Jamestown colony, England paid more attention to the attractive potential of the New World. In 1629, King Charles I granted Sir Robert Heath, his attorney general, a charter for a new colony, to be called Carolana, a Latin word meaning "Charles." The boundaries for Carolana were between the 31st and 36th parallel north. Although this overlapped the southern boundary of Virginia at the 34th parallel north, it did not cause a problem at the time because Sir Heath never promoted his land grant, and Carolana never existed.

During the English Civil War, Charles I was executed and his son escaped to Europe. After nine years of exile, Charles II returned to England to assume the throne, and in 1663, declaring that Sir Heath's charter had expired, he granted eight investors known as the Lords Proprietors with a charter for a new colony named Carolina in honor of his father. For the next hundred years, the energy, commitment and focus on the Carolina colony turned the tide for England's presence in North America and began its successful domination over the French and Spanish until the end of the Revolutionary War. The Carolina charter originally granted the Lords Proprietors the same area as Carolana, all the land from the 36th parallel north (near the Outer Banks of North Carolina) down to the 31st parallel north (near present-day Brunswick, Ga).

Not surprisingly, disputes about overlapping territory arose. Albemarle, a large estuary with rich wetlands and farm lands, was at the southern end of the Virginia colony, very close to the boundary set by Charles II, and soon colonists living in the area complained that their land was being taken over by the new Carolina colony. The Lords Proprietors, one of whom was a cousin to the king, responded to the Virginians' complaints by asking Charles for more land. In 1665, the charter was increased so that the northern boundary was 36° 30' north latitude and the southern boundary extended south to the 29th parallel. This, of course, included the Albemarle Sound area, which dismissed all the claims from settlers who had thought they owned land in Virginia. Additionally, the 29th parallel north (close to Daytona Beach, Fla.) is about 50 miles south of St. Augustine, and it did not take long for the Spanish to react with violence to the news that Carolina was now claiming ownership of land they had lived on for the past 100 years.

Settlers began pouring in to the vast Carolina, and two distinct areas developed. In the north, wealthy investors had huge plantations served by slaves from the West Indies. Charles Towne, in the south, became extremely successful because of its huge harbor, the center of a profitable shipping industry. Within about 20 years the Lords Proprietors organized separate governments for the two regions, and over the next 50 years the political governing units of each area grew stronger and more independent. Gradually the Lords Proprietors' control came under scrutiny, and in 1729, seven of them sold their interests in Carolina to the Crown. North and South Carolina were officially made royal colonies.

The Intercolonial Wars

The race to gain control in the New World resulted in ongoing bloodshed among English, Spanish and French colonies, which flared up as greater battles occurred in Europe. The European populace were relatively unconcerned about what was happening in the colonies since they had their own problems, but from the colonies' viewpoint, the European struggles ultimately determined the fate of the land on the North American continent. Four conflicts occurring over the course of about a hundred years are known collectively as the Intercolonial Wars.

In 1689 England joined the League of Augsburg, a coalition of European nations, to resist the invasion of the German Rhineland by France's King Louis XIV. The resulting War of the Grand Alliance rapidly crossed the Atlantic to involve the English and French colonists in New England, Montreal and Nova Scotia. The first of the Intercolonial Wars was known as King William's War. (Colonists named the Intercolonial Wars in honor of the English sovereign at the time of the conflict.) Both England and France had Native American allies; the Iroquois and Mohawk fought mainly with the French, but some tribes did fight alongside the English. A decade would pass before the European combatants met in the Dutch Republic and signed the Treaty of Ryswick, ending the European conflict in 1697. The treaty also ended King William's War in the colonies as a stalemate with no territorial acquisition by either side.

The European War of Spanish Succession began in 1702 over who would succeed the mentally and physically disabled Charles II of Spain. Both France and Austria claimed the throne because of complex intermarriages, and their respective allies joined the fray. The conflict spilled over into the New World and became known as Queen Anne's War, the second of the Intercolonial Wars. Another bloody decade saw settlements from Florida to Canada burned and thousands of people killed. The Treaty of Utrecht in 1713 ended the war in Europe and in the colonies, yet there was no resolution to the ever-present question of who would control North America.

The French were pushed out of Canada, now under English control, but still had numerous trading posts along the Mississippi River. The Spanish missions along the east coast of the continent were destroyed, but Spain remained firmly entrenched in Florida. England's colonies remained loyal to the Crown and continued to grow and trespass on native lands. And in another 30 years, the fighting among the Europeans would start again.

Native Americans

While colonization was driven by European ambitions, it must be acknowledged that hundreds of thousands of people had inhabited the New World for centuries prior to its "discovery." They attempted to protect and keep their land from Europeans, often resulting in decades of atrocities committed against their people and cultures. Spain established missions to convert the Native Americans to Catholicism, and the Spanish used forced labor and slavery. English colonists were also abusive to natives, sometimes partnering with one tribe to fight another, but always taking the property under the principle of "to the victor belong the spoils." On the other hand, the French lived with local inhabitants, adopted their traditions and language, and worked with them learning the fur trade; intermarriage was common and there was little or no effort to subjugate the people already on the land. Certainly, the indigenous populations played a significant role in the development of the North American continent and what would become the colony and eventually state of Georgia.

Important Dates I

1492 Christopher Columbus discovers the New World
1494 Treaty of Tordesillas between Spain and Portugal
1497 John Cabot claims Newfoundland/Nova Scotia for England
1513 Ponce de León claims Florida for Spain
1534 Jacques Cartier claims Canada (New France) for France
1539 De Soto's voyage to southern United States area
1565 St. Augustine founded by the Spanish
1583 Humphrey Gilbert claims Newfoundland for England
1585 First Roanoke colony settled
1607 Virginia Company of London formed
      Jamestown colony settled
1620 Plymouth colony settled
1629 Carolana charter granted by Charles I, never settled
1663 Carolina charter granted by Charles II and colony settled
1670 Charles Towne established
1689 All Spanish missions north of St. Marys River withdrawn
      King William's War
1702 Queen Anne's War

II

Georgia, the Thirteenth Colony

Although most attempts by English investors to establish colonial settlements during the 17th century had failed, the beginning of the 18th century found England in control of most of the Atlantic coast. By 1731 there were 12 separate geographical colonies along the eastern coastline with a total population of over 500,000 people and a recorded 1,500 transatlantic crossings. All of the colonies had developed a governing system modeled after Parliament, and many were directly under Crown rule, subject to English law and with a royal governor. (Pennsylvania and Maryland were proprietary colonies, and Connecticut and Rhode Island were chartered corporate entities; they also had a similar governing structure, but without direct English control.) All had significant religious freedom but also embraced slavery as a means for economic prosperity.

Even though the colonies were an economic drain, England was interested in growing its presence on the new continent and in removing any French and Spanish settlements already there. While five attempts to establish colonies would be made in the early 18th century, it would not be until 1732 that a group of investors and charitable organizers would be successful in persuading King George II to sponsor a 13th colony in the New World.

The five early attempts

The first attempt was in 1717 by Scotsman Sir Robert Montgomery, about whom little is known other than he was interested in settling colonies in Nova Scotia and the Carolinas. The Lords Proprietors of Carolina granted Montgomery lands south of Charles Towne for the establishment of a colony he called the Margravate of Azilia. ("Margravate" is a Germanic word for a territory ruled by the margrave, a military governor, but the meaning of "azilia" is unknown.) After several years of trying unsuccessfully to persuade new settlers to the region, Montgomery abandoned his project.

In 1726 Swiss merchant Jean Pierre Purry swayed the Lords Proprietors of Carolina to allow him to settle an area west of the Carolinas which he would call Georgia. Purry stressed the need for fortifications against the French who were coming down the Mississippi from Canada and spreading eastward to the Allegheny Mountains. In addition, he believed the latitudes of Georgia would be ideal to raise rice, silk and indigo. The Proprietors gave their approval, but a lack of funding proved to be too great an obstacle and the project was never begun.

The third attempt was by Thomas Coram, an Englishman who made his fortune as a ship builder in Massachusetts. In 1727 he proposed that a colony called Georgia, after England's new King George II, be settled on lands between Massachusetts and Nova Scotia as a refuge for Protestants from other European countries. Hemp would be the principal export. Massachusetts, however, did not want to give up any of its land and successfully lobbied the Crown to deny the proposal. Coram nonetheless made his mark on history by establishing the Foundling Hospital in London, the world's first home for abandoned children. The charity still exists as the Thomas Coram Foundation for Children.

When South Carolina became a royal colony in 1729, Governor Robert Johnson encouraged citizens to build settlements south of Charles Towne as a defense against the Spanish and Native Americans. Purry was again given permission for a new settlement, and by 1736 there were over 450 Swiss settlers in a village called Purrysburg on the South Carolina side of the Savannah River. Over the next decade, disease, hostilities with natives, and inadequate financing proved too much for the settlers, and the township failed. All that remains now is Purrysburg Landing, an access for boaters on the Savannah River.

In 1731 a fifth attempt for a new colony was made by Sir William Keith, the retired royal governor of Pennsylvania, and two other investors. They asked England's Board of Trade for a land grant of 120,000 square miles in what is now West Virginia, Ohio and Indiana for a new colony, which they also wanted to call Georgia. A paid militia would protect the colony while German and Swiss Protestants, skilled in the production of potash for fertilizer and saltpeter for gunpowder, would provide economic support. However, Virginia and Pennsylvania did not want the competition and complained to the Board of Trade which then denied Sir Keith's request.

The Trustees of Georgia

England during the 1700s was experiencing an Age of Enlightenment, a time of rethinking the old institutional ideas. Philosophical and intellectual developments brought a change of attitude regarding the common citizen. Charity, philanthropy and social reform were embraced as ideals. The timing was perfect for 30-year-old James Oglethorpe.

He was born in 1696 into a privileged family; his father was a member of Parliament and a general in the Royal Army of King James II. Oglethorpe was educated at Eton and Corpus Christi College of Oxford but left after two years to pursue a military career. After making a name for himself during a military campaign involving England and Turkey, he returned home and was elected a member of Parliament where he became a social reformer and launched a national campaign for prison reform. It was this zeal that brought him into contact with others of the same point of view.

In 1730 Oglethorpe convened a group of philanthropists, politicians, members of the clergy, and financial investors to discuss the possibility of establishing another colony in the New World. Again catering to King George II's vanity, they called their colony Georgia.

Well aware of the previous failed attempts to start a 13th colony, these Trustees of Georgia carefully planned their first step in the approval process, presenting their petition to Parliament's Board of Trade. They had already received tacit approval from officials in Charles Towne, which was becoming an important maritime commercial center for the profitable slave trade and commodities such as rice, rum, indigo and tobacco.

The Trustees also understood the need for a stronger military presence south and west of the Carolinas. The territory between the

Savannah River and St. Augustine, known as "the debatable lands," was an area of conflict with the Spanish coming north from Florida and the French coming east from Mississippi, both looking to trade with the Creek and Cherokee people. These struggles were interfering with the Crown's imperialism on the Atlantic coast, and a southern colony would provide a barrier to the other countries as well as showing England's strength.

The new colony

For the first time in 50 years and as a direct confrontation to the Spanish who claimed the same land, the king granted a new charter to the Trustees in 1732 providing for a new colony whose boundaries were

> ... all those lands, countrys and territories, situate, lying and being in that part of South-Carolina, in America, which lies from the most northern part of a stream or river there, commonly called the Savannah, all along the sea coast to the southward, unto the most southern stream of a certain other great water or river called the Alatamaha, and westerly from the heads of the said rivers respectively, in direct lines to the south seas; and all that share, circuit and precinct of land, within the said boundaries, with the islands on the sea, lying opposite to the eastern coast of the said lands, within twenty leagues of the same, which are not inhabited already, or settled by any authority derived from the crown of Great-Britain ... .

The Savannah River is 350 miles long and at the time was formed by the junction of the Tugaloo and Keowee rivers. Before entering the Atlantic Ocean, the river broadens into an estuary of channels which is rich with marine and plant life. Both Savannah and Augusta were founded on the river and were served commercially by sailing vessels in the 18th century and steamboats in the 19th century. The slow-moving Altamaha River winds almost 150 miles from its origin at the confluence of the Ocmulgee and Oconee rivers near Hazelhurst, Ga., to its mouth on the Atlantic Ocean about 20 miles north of Brunswick, Ga. The Altamaha River watershed is one of the three largest river basins on the Atlantic coastline, draining approximately one-quarter of the state of Georgia and discharging approximately

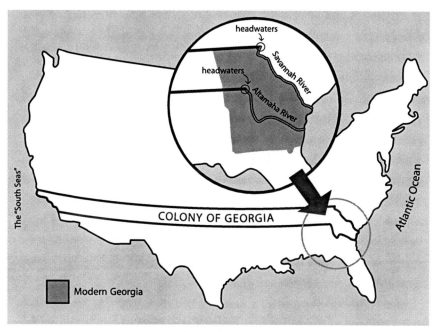

Trustees' charter boundaries, 1732

100,000 gallons of fresh water into the ocean every second.

The choice of the Altamaha River as the southern boundary of Georgia was not arbitrary. The Carolinians were concerned that the French would hinder their trading with the Native Americans and attempt to permanently settle lands west of the Alleghenies. As early as 1720, a report concerning South Carolina's defense system was sent to England's Board of Trade suggesting the headwaters of the Altamaha River as a location for a fort to prevent any French invasion. The Board of Trade quickly authorized the construction of the fort, and for the next 10 years, England and South Carolina promoted the Altamaha River area to attract settlers. The Trustees' plan for a new southern colony could not have come at a better time.

The Trustees' charter provided for different conditions than the charters of the previous 12 colonies. Georgia was the only proprietary colony to receive a stipend from England to help with its success and survival. The Trustees themselves were not allowed to have any administrative leadership position nor own any land in the colony. There was to be no importation of slaves or rum and, except for Catholics and Jews, religious freedom was guaranteed. The term of the charter was for 21 years, at which time it could be renewed or revert to the Crown.

While philanthropy and charity were the main motivations for establishing Georgia, the Trustees were also astute businessmen who knew that a strong economic environment was the key to Georgia's success. The other colonies were exporting rice, cotton, wine and tobacco back to England, but the Trustees had a plan which they believed would not only guarantee Georgia's survival but would also provide a profit for their investment. They believed that since the latitude of Georgia was the same as China, the climate would be perfect for raising mulberry trees and that silk could be produced cheaply and in such quantities that exporting silk to England would contribute to the survival of their enterprise.

Since Italy was the prime source of silk in Europe, several Italians were brought over on the first boat of colonists to teach the art of silk production. But the Italian experts soon discovered that the mulberry tree which grew in Georgia at that time was the red mulberry tree. The finicky silkworm will only eat the leaf of the white mulberry tree, so the Trustees imported hundreds of white mulberry trees, and each property owner was required to plant 100 trees on every ten acres of land. The men were taught how to grow the trees, and the women were taught how to raise the worms and wind the silk from the cocoons. Parliament exempted silk from any import tax, and by the mid-1750s over 10,000 pounds of raw silk was exported from Georgia to England annually. The silk was of such high quality that it sold for three times more than any silk imported from Italy or France.

Ultimately, all the colonies' silk industry declined because of the labor-intensive process of unwinding the silk thread from the cocoons. The decade preceding the Revolutionary War brought great demands on the average colonist, who became more focused on defeating the English rather than supplying them with any commodity.

In 1834 a group of businessmen attempted to revive the silk industry in Georgia. A large number of Chinese workers were settled in the North Georgia native village of Etowah, renamed Canton after the capital of China, to teach the local residents the art of sericulture. They were unsuccessful, and the Chinese population all but disappeared. Today, Canton (pronounced with the emphasis on the first syllable rather than the second) is a bedroom community of Atlanta with a thriving and growing economy.

The depiction of the life cycle of the silkworm on the Colonial Seal of Georgia, used for legitimizing legislative acts and legal documents, emphasizes the importance of silk production to the new colony. On the obverse of the seal, two figures rest on urns with water pouring out of them, representing the Altamaha and Savannah rivers. Between them is another figure, the Genius of the Colony, with a cap on her head representing liberty. A spear in one hand represents strength; her other hand holds a cornucopia which represents abundance. The words in Latin are often translated as "May the Colony of Georgia flourish." The reverse side of the seal depicts a silkworm and a cocoon on a mulberry leaf. It bears the famous motto *Non sibi sed allis,* meaning "Not for self, but for others," a testament to the tenets which inspired the Trustees to form the colony.

The original seal is housed in the British Museum in London; a facsimile is on prominent display at the Georgia Archives Building in Morrow, Ga.

Although the Trustees were prominent citizens who were interested in philanthropy and in providing religious freedoms, the foremost requirement for the initial settlers was military experience. In fact, Oglethorpe was the only trustee to ever visit the colony. After the royal charter was granted, he sailed to the New World with 35 families of volunteer merchants, cooks, carpenters and farmers on the adventure to settle Georgia. For the next 10 years the whole project relied on him, and he took command of everything from allocating land to erecting forts to building roads and bridges. His energy and commitment to the idea of starting Georgia is laudable, but his domineering personality created the schism which doomed the struggling new colony.

On Jan. 13, 1733, eight weeks after leaving England, the 200-ton frigate *Anne* landed near Charles Towne. Oglethorpe went ashore and received greetings, advice and supplies from Governor Johnson. The next day, the *Anne* sailed down to Port Royal on the island of Hilton Head, and the settlers were allowed to come ashore.

While it may be widely believed that Georgia was founded as a debtor's colony, none of the original settlers had been in debtor's prison. During the life of the Trustees' control of the colony, the number of debtors in Georgia was less than the other 12 colonies' debtor population. Some of the original colonists were called charity settlers, given passage and subsistence in exchange for certain types of work, but a larger number of colonists were the adventurer settlers who paid their own way and bought large land tracts.

Within a few days of landing, Oglethorpe and a small party of men sailed up the Savannah River to meet with Tomochichi, the chief of a small tribe of Creek known as the Yamacraw, to get permission to build a new settlement overlooking the Savannah River. The Creek had been dealing with Europeans for over 150 years and were wary of them, but the Yamacraw had very good relations with the Carolinians so Oglethorpe's meeting with Tomochichi went very well.

Relations with the Creek

The interpreter for that first and many subsequent meetings with the Native Americans was Mary Musgrove, the daughter of a Creek woman and an English trader from South Carolina. Musgrove spoke both English and Creek and was extremely valuable to maintaining good relations between the tribes and the new colonists for the next 15 years.

Chief Tomochichi was in his 90s at the time of his meeting with Oglethorpe and had the reputation of being a conscientious, intelligent leader. The two men had mutual respect for each other and remained friends for the rest of Tomochichi's life. When Oglethorpe returned to England in June 1734 to report to the Trustees of Georgia, he took the chief, his wife and his nephew. For the next several months, the Yamacraw were presented to the Trustees and to Parliament, where Tomochichi made a speech about friendship and peace. Oglethorpe also presented them to King George II, where the chief spoke about needing better education, trade and peace for his people. Toonahowi, Tomochichi's 15-year-old nephew, rapidly learned to speak English and became a close friend of the 15-year-old son of King George II. All of this positive diplomacy laid the foundation for years of excellent relations between the Creek and the Georgia colonists.

A few years after returning from England, Oglethorpe, Tomochichi and Toonahowi were exploring the barrier islands off the coast south of Savannah looking for a site to build a fort. Toonahowi suggested that they rename Isla San Pedro, named by the Spanish as part of their mission chain but now within the Creek Nation boundaries, in honor of King George's son, William, Duke of Cumberland. Today, Cumberland Island, located at the boundary between Georgia and Florida, is part of the National Park Service and visited by thousands of tourists every year.

Tomochichi died in October 1739, at age 97, after a brief illness. After being given a military funeral by Oglethorpe, who was one of his pallbearers, he was buried in Savannah where his tomb is marked by a piece of granite taken from Stone Mountain, Ga. Toonahowi became chief of the Yamacraw tribe and remained a strong ally of the English until he was killed in battle in 1743.

First settlements

In 1734, 500 additional settlers came to Georgia, and within a few years additional townships were settled surrounding Savannah as part of the defensive posture so important to Oglethorpe.

He recognized that building a fort on the Georgia side of the Savannah River across from a village called Savannah Towne, which had been a trading post for over 60 years on the Carolina eastern side of the river about 150 miles upstream from Savannah, would be a strong deterrent to the French coming in from the west and, at the same time, would make trading easier with the Creek who lived on the western side of the river. He convinced some traders to leave Savannah Towne and start a new village, which he called Augusta in honor of the daughter-in-law of King George II.

Augusta continued to thrive during the colonial period and became the capital of Georgia during the American Revolution. The village of Savannah Towne in South Carolina, however, slowly declined and within 20 years it disappeared entirely; 150 years later, a new village called Aiken, after William Aiken, a prominent Charleston banker, was built on the same land as part of a new railroad system and served as a winter resort for the wealthy.

Darien, a settlement 60 miles south of Savannah at the mouth of the Altamaha River, was settled by Scottish immigrants in 1736. The fiercely proud and well-trained Scots proved to be a vital part of the militia providing protection against the Spanish coming north from Florida.

The village of Ebenezer, only 25 miles north of Savannah on the Savannah River, was settled as a religious refuge for Lutherans looking for relief from persecution in their native Austria. They were successful merchants and farmers but were not involved with the other Georgia colonists because of their refusal to learn English. The Salzburgers, as they were called, played a prominent role during the Trustees' tenure in Georgia because of their belief in religious free-

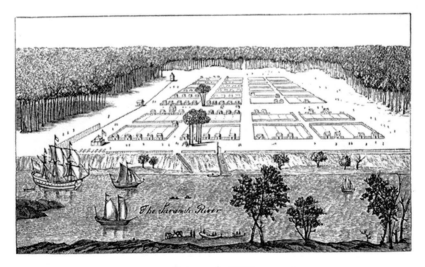

Savannah, 1741

dom, no slaves and no rum. Eventually more than 1,500 Salzburgers settled in the area and became the most successful of all Georgia colonists to grow the mulberry trees needed for the silk industry.

Dissent and the Malcontents

Although Georgia's charter provided that none of the Trustees could hold any administrative leadership position in the colony, Oglethorpe assumed command and never established a governing body to manage the affairs of the colony. Without an entity controlling the citizens, there was almost complete lawlessness. There were no police or attorneys, only Oglethorpe as leader. A group of men, calling themselves the Malcontents, began meeting weekly at one of the taverns to voice their concerns. Complaining to Oglethorpe did no good, so eventually the group petitioned the Trustees in England for a change in Georgia's administration. When the Trustees ignored the petition, many of the group's leaders moved from Georgia to Charles Towne, where they continued their pursuit of demanding change in Georgia.

In 1737 the Trustees hired William Stephens as the secretary of the colony believing that his appointment would mitigate Oglethorpe's poor administrative skills, although at first Stephens had no authority to act without Oglethorpe's permission.

Stephens wrote a paper in 1740 entitled "A State of the Province of Georgia." The paper claimed that the policies and procedures of the Trustees enjoyed great support throughout Georgia and that the colony was an economic success. One of the Malcontents in Charles Towne immediately rebutted Stephens' paper with "A True and Historical Narrative of the Colony of Georgia." Both papers were circulated in the colonies and England but, even though both papers were widely read, given the Trustees' considerable economic and political power, no real changes occurred in Georgia.

In response to his father's paper, Thomas Stephens, one of the original founders of the Malcontents, also published his own widely circulated paper entitled "The Hard Case of the Distressed People of Georgia." Thomas made several trips to England, appearing before Parliament to make his case against the Trustees. Although he achieved no immediate success, his efforts were rewarded a few years later when the restrictions on rum and land ownership were lifted.

In 1741 the Trustees appointed the elder Stephens, who had developed an increasingly antagonistic relationship with Oglethorpe, as the president of the colony in addition to being secretary. (The title "president" was used, rather than governor, to avoid seeking permission from the king.) Oglethorpe's power and control were diluted. He was told to look after military affairs, while the Trustees and Stephens would look after the colony. Oglethorpe, who probably suspected something like this would happen, left Savannah almost immediately and took a military regiment with him to build Fort Frederica on St. Simons Island. While Stephens never achieved the popularity of Oglethorpe, he administered the affairs of the colony well and served as Georgia's president until 1751.

The War of Jenkins' Ear and Oglethorpe
The colonies enjoyed a relatively quiet period after the end of the Queen Anne's War in 1714, but sporadic conflicts continued between England and Spain in the Caribbean and West Indies. In 1731 during a naval battle near Havana, Cuba, the victorious Spanish commander boarded the English brig *Rebecca*, commanded by Captain Robert Jenkins. Captain Julio de León Fandiño cut off Jenkins' ear, giving it back to him with the epithet, "Were the King of England here and also in violation of the laws, I would do the same for him!" England and Spain tried to settle their differences, but the meetings only served to increase the animosity between the two nations. In 1738 Jenkins appeared before the House of Commons in London, waving the jar in which he had kept his severed ear. His testimony about the Spanish so incensed the members, as well as the public, that war was declared on Spain in 1739.

The War of Jenkins' Ear exploded in the southern colonies. For three years the Georgia-based Royal Militia made thrusts into Florida as hostilities increased over the debatable lands. Oglethorpe and his militia, with additional troops provided by England, invaded Florida and attempted to lay siege to St. Augustine in the summer of 1740 but were unsuccessful. Two years later, the Spanish launched a full-scale naval and land battle to gain control of St. Simons Island. Oglethorpe withdrew from Fort St. Simon before the Spanish could attack and concentrated his forces at Fort Frederica, of which the Spanish were unaware. Following a defense planned eight years earlier, he carefully placed and rallied his troops and succeeded in soundly defeating the Spanish in what is known as the Battle of Bloody Marsh. This decisive rout ended the War of Jenkins' Ear, and Georgians commemorate this event at Wormslow Plantation in Savannah on the fourth Saturday of every May.

More than anything else Oglethorpe accomplished in Georgia, his defeat of the Spanish cemented his reputation as a military hero. His foresight in building Fort Frederica proved to be pivotal both in his life as well as in the survival of all of England's colonies. If the Spanish had defeated Oglethorpe at Bloody Marsh, historians believe that Spanish armada and ground forces would have continued up the coast toward Charles Towne and the Carolinas with significant success. After the Battle of Bloody Marsh, Spain was never again a threat to the English on the Atlantic coast.

Within a year, Oglethorpe left to face a court-martial when one of his officers made a formal complaint to London about him. After an extremely short investigation, he was cleared of all charges and declared a hero. Parliament also repaid him the money that he had loaned to the colony over the preceding decade. Oglethorpe never returned to Georgia or North America although he felt a connection to the colony for the rest of his life.

In 1744 the 48-year-old bachelor married Elizabeth Wright, heir to the huge estate of Cranham. He lost his Parliamentary seat in 1752 but was elected a fellow of the Royal Society and was a also founding member of the British Museum. By age 65 he had become a gentleman scholar and a man of letters; he had an extensive library, read and quoted the classics in Latin, and developed a following of such notables as James Boswell, Samuel Johnson and Sir Joshua Reynolds, who painted his portrait. Oglethorpe died in 1785 at age 89.

Georgia and the United States have paid homage to Oglethorpe in many ways. Savannah has numerous plaques, granite monuments, statues, squares and streets named after him. Numerous clubs, buildings, schools, banks and hotels have also borne his name. There is an Oglethorpe County in Georgia, a town of Oglethorpe in middle Georgia, and Oglethorpe University located in Atlanta. The U.S. Postal Service has issued two stamps with Oglethorpe's portrait on them: one in 1933 recognizing Georgia's bicentennial, the second in 1983 on Georgia's 250th anniversary.

The Intercolonial Wars continue

The War of Austrian Succession began in Europe in 1740 and spread to the colonies as King George's War, the third Intercolonial War. Most of the fighting involved northern New England and Nova Scotia. The English captured Louisbourg, the French fortress on Cape Breton, Nova Scotia, which gave them control over all of Canada. In 1748, after almost a decade of fighting, the Treaty of Aix-la-Chapelle ended the European war as well as King George's War. Louisbourg was returned to the French, much to the chagrin of the English colonists.

For the next decade both France and England aggressively pursued their policy of claiming land in North America to further their economic interests. Both nations built forts and trading posts and claimed lands from west of the Appalachian Mountains to the Mississippi River and from the Great Lakes to the Gulf of Mexico. Both courted the Native Americans who lived in those areas, hoping for their allegiance in the event of another armed conflict. Eventually the conflicting goals of the two powers would explode into a bloody war.

Often seen as a continuation of the War of the Austrian Succession, the Seven Years' War began in 1756 and caused over a million deaths, massive destruction of property, and huge changes in the political map. Again Europe's war spilled over into the colonies, where it was known as the French and Indian War. (Since there had already been a King George's War, this conflict was named for England's opponents, the French and the Indians.) The battles raged in Europe, Africa, India, North America, the Caribbean and the Philippines. Winston Churchill called it "the first world war in human history" as it was the first conflict to be fought in territories around the globe.

The French and Indian War was the last and most important of the Intercolonial Wars because of its result. In his book *The Scratch of a Pen: 1763 and the Transformation of North America*, Colin Calloway states, "After the Treaty of Paris was signed in 1763, more American territory changed hands than in any treaty before or since." This treaty provided that France give up all the territory called New France to England.

> In order to re-establish peace on solid and durable foundations, and to remove for ever all subject of dispute with regard to the limits of the British and French territories on the continent of America; it is agreed, that, for the future, the confines between the dominions of his Britannick Majesty and those of his Most Christian Majesty, in that part of the world, shall be fixed irrevocably by a line drawn along the middle of the River Mississippi, from its source to the river Iberville, and from thence, by a line drawn along the middle of this river, and the lakes Maurepas and Pontchartrain to the sea; and for this purpose, the Most Christian King cedes in full right, and guaranties to his Britannick Majesty the river and port of the Mobile, and every thing which he possesses, or ought to possess, on the left side of the river Mississippi, except the town of New Orleans and the island in which it is situated, which shall remain to France, provided that the navigation of the river Mississippi shall be equally free, as well to the subjects of Great Britain as to those of France, in its whole breadth and length, from its source to the sea … .

In other words, England now controlled all the land east of the Mississippi River. And in a secret treaty during the last months of the war, France also ceded Louisiana, its territory west of the Mississippi River, to Spain.

France would never again have any significant presence on the North American continent. The reality was that the French citizens in Quebec and Montreal, as well as the Acadians from Nova Scotia who moved down the Mississippi River to New Orleans, could be loyal to any controlling government. They started French trading posts and villages throughout North America, and their language and culture remained French as they intermarried with the local populations.

The Treaty of Paris also provided that Spain would lose all of its missions and forts in Florida and in return be given the valuable island of Cuba. Louisiana now belonged to Spain as well, but because of the huge costs involved, it would take six years for Spain to begin administrative and military control of the territory.

Interestingly enough, no representatives of the Native Americans played any role in the discussions. They were ignored, and England planned to deal with each tribe separately as the need arose.

Although the colony of Georgia was not mentioned in the treaty, its boundaries were nonetheless affected. Since England had given up its territory west of the Mississippi, Georgia's western boundary was no longer the "south seas" but now stopped at the Mississippi River, placating France and Spain who had their own agenda exploring the lands beckoning further to the west. The debatable lands to the south of Georgia no longer existed.

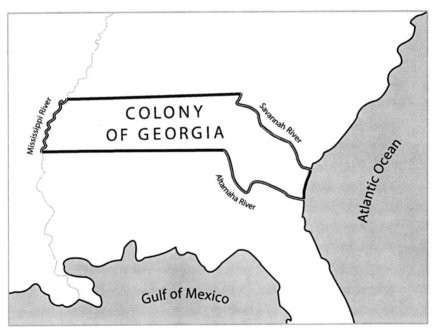

After the Treaty of Paris, 1763

The Royal Period of Georgia

In the late 1740s, after massive crop failures and the ever-present

Native American hostilities, less than 300 settlers remained in Savannah. The rest of the population had moved north to the Carolinas and Virginia, both of which had good economic opportunities and a representative government. By 1750 there were only 1,500 persons living in the few towns of Georgia, almost 500 of which were slaves, along with about 700 royal militia stationed in southern Georgia.

After years of struggling with financial problems, the Trustees surrendered their colony to King George II one year before their charter was to expire. The years from 1752 until 1776, when the colony declared its independence from England, are referred to as the Royal Period. The Board of Trade sent a temporary president to form an interim government, the Provincial Council, and the remaining citizens welcomed the new authority because of the need for law and order. The old Trustee mandate of "no slaves and no rum" was ignored, and immediately the economy improved. The boundaries of the royal colony of Georgia remained the same as what the Trustees had been granted.

In 1754 the first of three royal governors came to Georgia. John Reynolds, an English Navy captain, instituted a strong legislature and a desperately needed court system in the colony. Within two years, the population of the colony swelled to approximately 9,000 inhabitants including at least 1,000 slaves. Shortly after arriving in Savannah, Reynolds wrote the Board of Trade that,

> I must be leave to observe that the bounds of this province, as expressed in my commission, is so uncertain with respect to the southern boundary, that great inconveniences will attend it; for nobody in this country knows where the head of the Alatamaha [sic] River lies, and they are all as ignorant of what may be called the most southern stream of it ... .

Reynolds proposed that a line drawn from the mouth of the Altamaha River at the Atlantic Ocean due west to the South Seas be the southern boundary of Georgia. This differed from the boundary line described in the Royal Charter that the westward line start from the head of the Altamaha River. But despite continued appeals over the next three years, there was no response from the Board of Trade about the new boundary.

It soon became apparent that Reynold's military training and experience did not provide him with the political skills necessary to institute a new government. His unsuccessful diplomacy with the Creek caused serious conflicts in Georgia, and after three years he was recalled to England.

Henry Ellis, the second royal governor, reiterated the need for a new southern boundary of Georgia. The problem over the debatable lands south of the Altamaha River had subsided ever since the Battle of Bloody Marsh 15 years before, and the Spanish now considered the northern boundary of Florida to be 60 miles south of the St. Marys River. (According to the U.S. Board on Geographic Names, there is no apostrophe in "Marys.") Ellis requested the Board of Trade extend Georgia's southern boundary by 100 miles, but this request was denied.

Ellis is best remembered as a skilled diplomat and politician. He developed a budget and taxes for the colony's expenditures, made peace with the Creek, and gave Georgians a new sense of pride. Beginning in 1758, Georgia was divided into 12 social, religious and political districts known as parishes, which elected delegates to represent them in Georgia's Royal General Assembly. Unfortunately poor health cut his time in Georgia short, and after three years Ellis resigned his position and returned to England.

James Wright was the third and last of the royal governors of Georgia. Although born in England, he came to South Carolina at age 14 when his father was appointed chief justice. Wright also practiced law and became the attorney general of South Carolina. For a few years he represented South Carolina as its agent in London until he was appointed lieutenant governor of Georgia under Ellis in 1757. Wright owned thousands of acres of land, employed over 500 slaves, and was a successful plantation owner. He had married in South Carolina and had six children, although his wife died shortly after their coming to Georgia. When Ellis resigned in 1760, Wright succeeded him and proved to be the most popular and respected governor of all the colonies.

Wright was governor after the French and Indian War when England owned all the land in the New World east of the Mississippi.

Under his leadership, Georgia's economy thrived and relationships with the Native Americans improved. The colony's new boundaries included a huge amount of land inhabited by the Creek, and Wright negotiated treaties with them increasing Georgia's western lands from one million to six million acres by the end of his tenure. He provided a stable government for Georgians who remained loyal to the Crown, but as the seeds of revolution gradually came to Georgia, his control over the colony eroded until he was forced to leave during the early years of the Revolutionary War.

After only 30 years of existence, Georgia's territory had significantly decreased, but since no one had ever explored much of the interior of the New World, the new boundary had no appreciable effect. The next several years of England's control over the Atlantic coastline would bring more dramatic changes to Georgia's boundaries.

Important Dates II

- 1708   Carolina formally separated into North and South
- 1717   Sir Robert Montgomery's Margravate of Azilia, never settled
- 1726   Jean Pierre Purry's Georgia colony, never settled
- 1727   Thomas Coram's Georgia colony, denied
- 1729   Lords Proprietors give up their charter;
  North and South Carolina become royal colonies

  Purrysburg settlement, unsuccessful
- 1730   Trustees of Georgia organize
- 1731   Sir William Keith's Georgia colony, denied
- 1732   Charter of Georgia granted by George II
- 1733   James Oglethorpe and colonists land at Savannah
- 1735   Malcontents organize
- 1737   William Stephens appointed secretary of Georgia
- 1739   War of Jenkins' Ear
- 1740   James Oglethorpe invades Spanish Florida

  King George's War
- 1741   William Stephens appointed president of Georgia
- 1742   Battle of Bloody Marsh
- 1743   James Oglethorpe leaves Georgia
- 1752   Trustees of Georgia surrender charter; Royal Period begins

  Gregorian calendar observed in England and colonies
- 1754   First royal governor arrives in Georgia
- 1756   Seven Years' War fought around the world;
  French and Indian War begins in the colonies
- 1760   George II dies, succeeded by his grandson George III
- 1763   Treaty of Paris ends French and Indian War

III

The Birth of a State and a Nation

Georgia played a very different role from the other colonies during the rebellion years. Geographically and politically, it was a long way from the epicenter of the social and political unrest.

When Oglethorpe's settlers came to Georgia, they were the first new colonists on the eastern coast for over 50 years, and they were very connected and loyal to England. After the Trustees gave up their control over the failing colony, the success of the three royal governors furthered Georgia's continued loyalty to the Crown. In its 20 years as a royal colony, Georgia had grown to a population of 35,000 including almost 17,000 slaves. Plantations of over a thousand acres were common, and there was a good system of roads along the coast and up to Augusta. Savannah, Darien and Brunswick were busy ports contributing to the overall rising economy. Georgia had strong family ties to England, whereas the older colonies, some of which had been geographically isolated from the motherland for over a hundred years, did not have those bonds. Many of the first generation born in Georgia held the same loyal feelings to England as their parents. The colony also had an excellent economy based on strong exports of cotton, wheat, silk and lumber. With their increasing wealth, Georgians were thinking about better education and a higher standard of living — not rebellion. Voting in Georgia could bring results, and while Georgia had its own militia, she also enjoyed protection from the native populations by the English military, whereas the other colonies viewed the redcoats as a domineering presence and not one of protection.

Little by little, however, current events began shaping the policy of resistance in Georgia.

The Proclamation Line and the Florida colonies

King George II died in 1760 and was succeeded by his grandson George III, the first English monarch of the royal dynasty of the House of Hanover to be born in England and speak English as his native language. He would rule England for 60 years, the second-longest reign in English history.

In October 1763, some eight months after the Treaty of Paris was signed, George III promulgated the Proclamation Act of 1763 regarding England's new ownership of land in North America. Along with land grants given to men who served in the military during the war, it created three new colonies — Quebec, East Florida and West Florida — in the New World, and a fourth colony — Grenada — in the West Indies. The Proclamation Line was also intended as an olive branch to the Native Americans, to appease them and avoid further bloodshed. Many tribes had allied themselves more with the French rather than the English because the French were interested in trading with them while the English were intent on taking their land.

The most significant and extremely unpopular part of the Proclamation Act to the colonists was that the Crown reserved to itself "all lands not included within the boundaries of the thirteen colonies and all the land lying to the Westward of the Sources of the Rivers which fall into the Sea from the West and North West." This obscure language was interpreted to mean that all the land west of the heads of all the rivers on the new continent which flowed to the

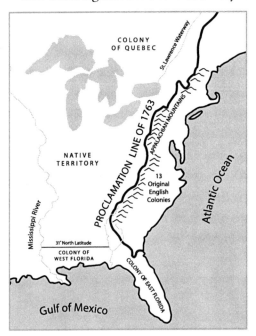

Proclamation Line of 1763

Atlantic Ocean was owned by the Crown and was off-limits to the colonists. Maps were drawn and disseminated to the colonists showing the Proclamation Line extending from Canada southerly on the western side of the Appalachian Mountains behind Georgia and then curving southeasterly down to the mouth of the St. Marys River at the Atlantic Ocean.

No colonist was allowed to own, live on or buy any of this land without a special license from the Crown. Numerous English military forts were built along the Proclamation Line, more to prevent the colonists from moving into those territories than to protect the public. The Proclamation Line would be an important factor in the alienation between the colonies and the Crown.

The new Florida colonies were described by familiar natural landmarks. West Florida was

> ... bounded to the Southward by the Gulph of Mexico, including all Islands within Six Leagues of the Coast, from the River Apalachicola to Lake Pontchartrain; to the Westward by the said Lake, the Lake Maurepas, and the River Mississippi; to the Northward by a Line drawn due East from that part of the River Mississippi which lies in 31 Degrees North Latitude, to the River Apalachicola or Chatahouchee; and to the Eastward by the said River.

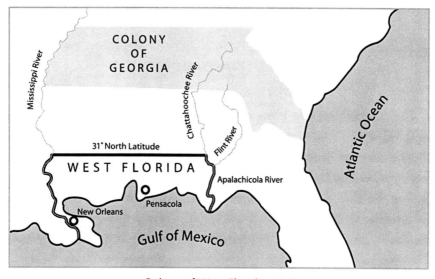

Colony of West Florida, 1763

East Florida was described as

> ... bounded to the Westward by the Gulph of Mexico and the Apalachicola River; to the Northward by a Line drawn from that part of the said River where the Chatahouchee and Flint Rivers meet, to the source of St. Mary's River, and by the course of the said River to the Atlantic Ocean; and to the Eastward and Southward by the Atlantic Ocean and the Gulph of Florida, including all Islands within Six Leagues of the Sea Coast.

Colony of East Florida, 1763

The Chattahoochee, Flint and St. Marys rivers were well known and important to the commerce of the Native Americans as well as the colonies of South Carolina and Georgia. Cherokee, Creek and other tribes had built villages along the Chattahoochee River long before the Europeans arrived, all of which may have had thousands of inhabitants during the 16th and 17th century. The Flint River begins as groundwater seepage from fractured rocks in north central Georgia (a victim of urban sprawl, the headwaters have now been diverted through a concrete culvert under the runways of Hartsfield-

Jackson Atlanta International Airport) and flows in a convoluted 220-mile southwesterly course to meet the Chattahoochee and create the Apalachicola River, which flows for a hundred miles into the Gulf of Mexico. These three rivers were navigable and important commercially until the late 1800s when shifting riverbeds and land accretion made the rivers unsafe. The St. Marys River, first named by Jean Ribault, a Spanish explorer of the region in the mid-1500s, was used by the Spanish and English for almost 200 years to transport cotton, sugar cane, rice and timber, including pine logs used for masts on Royal Navy ships; then steamers carried passengers, cargo and mail until the early 1900s when the lumber industry ceased.

At the request of Governor Wright, the Board of Trade by royal proclamation in January 1764 enlarged Georgia to encompass the adjacent territory which had not been granted to the Florida colonies. The colony was now

> ... bounded on the North by the most northern Stream of a River there commonly called Savannah as far as the Head of the said River and from thence Westward as far as his Majesty's Territories extended, on the East by the Sea Coast from the said River Savannah to the most Southern Stream of a certain other River called St. Mary, includ-

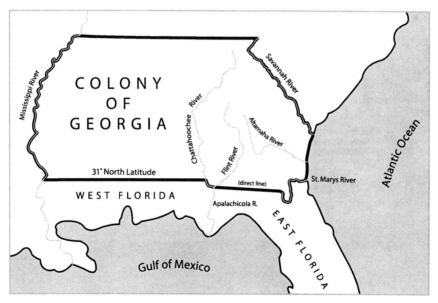

Result of Gov. Wright's request to Board of Trade, 1764

ing all Islands within twenty Leagues of the Coast lying between the said River Savannah and St. Mary, and on the South by the said River St. Mary as far as the Head thereof, and from thence Westward as far as his Majesty's Territories extend by the North boundary Line of his Province of East and West Florida.

These new boundaries gave a tremendous amount of land to Georgia, most of which was occupied by different native peoples and remained unexplored and unsettled.

After the West Florida colony was established, Governor General George Johnstone of West Florida wanted to expand the boundary of his colony northward to include land alongside the Mississippi and Chattahoochee rivers where royal trading posts were located, and in 1767, the Crown issued a supplement moving the northern boundary of West Florida to a line drawn from the confluence of the Yazoo and Mississippi rivers eastward to the Chattahoochee. This location is at Vicksburg, Miss., approximately 32° 28' latitude north, pushing West Florida almost 100 miles northward into Georgia's territory.

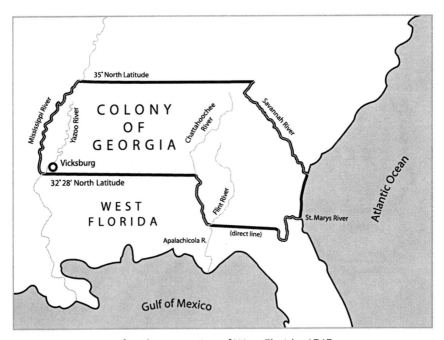

After the expansion of West Florida, 1767

Unrest in the colonies

An unintended by-product of the Intercolonial Wars was that the English colonies developed mutual trust, interdependence and communication. They gained an awareness of their own political and military strength and, eventually, their own unique identity of commonality of defense against the native populations, intercolonial commerce and American — not English — cultural ties. As the colonies became more interdependent, their frustration with the draconian English rule increased.

The years of malcontent in the original colonies on the Atlantic coastline began shortly after the end of the French and Indian War in 1763 when England passed a series of revenue-raising laws. The economy of the colonies included exports of silk, furs, rice, and tobacco back to London, and they depended on England and other European countries for imports. Despite their loyalty to the Crown for the past 150 years, burdensome import and export taxes were now being imposed.

The Sugar Act of April 1764 replaced the 30-year-old and largely ignored Sugar and Molasses Act to discourage the importation of molasses from the French, Spanish and Dutch West Indies. This new Sugar Act had lower taxes, more strict monitoring and penalties, and provided that lumber and other goods could only be exported to England.

In 1765 a Stamp Act placed a tax on all legal documents such as wills, contracts, permits and even newspapers. This was not an import tax but was instead a tax aimed, for the first time, directly at the colonies. It was met with significant opposition in written and vocal protests. The only colony to collect the tax was Georgia, whose citizens were much more loyal to the Crown than the other colonies, whose citizens included third- and fourth-generation Americans with no knowledge of England and very little loyalty.

Parliament also passed the Quartering Act in 1765, which required the public provide housing, food, alcohol and bedding to the English military. This was similar to a previous Quartering Act in force during the Intercolonial Wars, but now there was no war, and the colonists felt there was no need for the constant show of English military power nor the requirement that they provide for the troops. The revenue-raising laws and the Quartering Act served as the basis for the colonist's frequent cry of "no taxation without representation." For decades, the colonists, some 3,500 miles from England, were estranged from the very government which had control over their lives and the idea against such control was growing.

In 1766 Parliament repealed both the Sugar and Stamp acts after sensing significant opposition from the colonies. But to make a point, Parliament then passed the Declaratory Act stating that England had the power and authority to pass any acts necessary for the "common good."

Next to enrage the colonists was the Townshend Acts of 1767 which included an import tax on paper, paint, glass and tea, among other commodities. In addition, the act provided for "writs of assistance" which allowed tax collectors to search for smuggled goods. Smugglers, many of whom were merchants and ship owners, enjoyed a good living smuggling in highly taxed goods, and the ability of English tax collectors to arbitrarily search their ships was another English affront. Both John Hancock and Samuel Adams, later to be signers of the Declaration of Independence, were prominent in the smuggling business. Again, an uproar came from the colonists which resulted in parts of the Townshend Acts being repealed, but the taxes on tea and other commodities remained.

In late 1772 Massachusetts, followed by Virginia a few months later, formed a Committee of Correspondence to coordinate and disseminate written communication among the colonies. Similar to an underground newspaper, the colonial committees rallied opposition on common causes and established plans for collective action. Many committee members were also members of the colonial legislative assemblies and were active in the Sons of Liberty, a secret society organized for more violent revenge. Georgia was one of the first of the colonies to establish both a Committee of Correspondence and the Sons of Liberty. In Georgia, the Sons of Liberty were more open than in the other colonies and did not endorse the idea of physical violence. The members were more interested in preserving self-government, trading with England, and broadening Georgia's commerce with other nations.

The East India Trading Company, granted a royal charter in 1600, originally did business in India and eventually spread to the Caribbean islands. It became so huge that any threat to it was viewed as a threat to England. In 1773 Parliament passed the Tea Act, which gave the company a chance to avert bankruptcy by granting a monopoly on the importation of tea into the colonies. The new regulations allowed the company to sell tea to the colonists at a price lower than the price of smuggled tea, even including the required duty. This so angered the colonists that they boycotted any tea the company brought into the colonies. In December some Massachusetts citizens and Sons of Liberty, dressed as Mohawk, threw hundreds of crates of tea worth $2 million off ships anchored in Boston harbor.

In response to the Boston Tea Party, England shut down the harbor until reparations were made to the East India Trading Company. In addition, England passed the Coercive Acts, known to the colonists as the Intolerable Acts, aimed at weakening Massachusetts' legislature while strengthening the Crown's authority. These laws made it difficult for moderates in the colonies to speak in favor of Parliament, but it also unified the other colonies in their sympathy for Massachusetts.

The path to independence

Seeking to right the wrongs that had been inflicted on the colonies and hoping that a unified voice would gain them a hearing in London, the First Continental Congress assembled in Philadelphia in the fall of 1774. All the colonies on the North American continent except Georgia, East and West Florida and Quebec sent delegates. Among the 56 men were George Washington, John Jay, John Adams, John Hancock and Patrick Henry. The First Continental Congress had three objectives: to compose a statement of colonial rights, to identify Parliament's violation of the rights, and to provide a plan that would convince England to restore these rights.

The congressional delegates petitioned King George III in a document entitled "Declaration and Resolves of the First Continental

Congress" with the idea of reconciliation and firmness, but not of independence. While the delegates were probably the most radical of mainstream colonists, they were still loyalists and not insurgents. Words and phrases such as "life, liberty and property" and "the right to assemble" were used, but clearly the delegates were seeking a way to redress their grievances with the Crown while giving the colonies permission to govern themselves. The Congress also adopted a Continental Association wherein the delegates agreed to boycott English imports. The First Continental Congress lasted for two months, and the delegates agreed to reconvene in the spring if circumstances warranted.

On Jan. 17, 1775, Governor Wright appealed to the members of the Royal General Assembly to act as good citizens, not as radicals. His plea had little effect, however, because the very next day, more interested citizens also gathered in Savannah in what they called the First Provincial Congress to discuss whether Georgia should join the 12 other colonies who had agreed to prohibit importation and exportation of English goods. Only five of Georgia's parishes sent representatives to this meeting, but in spite of the small attendance, the representatives elected delegates for the Second Continental Congress.

The representatives also voted to ask the Royal General Assembly for approval of their actions, but Wright, rather than risking a vote on approving the policies of this First Provincial Congress, refused to convene a meeting. Events in the colonies moved very rapidly after that, and Parliament declared the American colonies in state of rebellion on Feb. 9, 1775. Georgia's Royal General Assembly would not meet again.

The delegates selected to attend the Second Continental Congress in Philadelphia refused to go because not all of the colony's parishes attended the First Provincial Congress. Instead, St. John's Parish, originally founded by English Puritans who migrated there from South Carolina, took an early stand for independence and voted to send Lyman Hall to the Second Continental Congress as a representative of the parish rather than the colony of Georgia. Hall was born in Connecticut and was a Yale-trained physician and minister. After a stormy confrontation with his parishioners, he moved to Georgia, practiced medicine, and became very prominent in revolutionary activities.

During that year, all the colonies were communicating with each other regarding their common complaints. Newspapers printed editorials, local protest groups met, and parishioners heard sermons about the need for the colonists to be more self-reliant and independent. The majority of Georgia's colonists were slow to consider any change in the political structure, but the battles of Lexington and Concord in April were pivotal conflicts that quickly changed the thinking.

Seven hundred men of the Royal Regulars in Boston marched the 20 miles to Concord to destroy the depot of the Massachusetts Militia and arrest John Hancock and John Adams. Word spread amongst the colonists, and a brief lopsided clash occurred in Lexington between the Regulars and a few dozen colonists. The skirmish left eight colonists dead and many wounded. The Regulars continued on to Concord, six miles away, where they were met by 300 of the Massachusetts Militia. The well-trained Royal Regulars easily overwhelmed the militia and continued with their raid on the Massachusetts depot, but when the Regulars began marching back to Boston, they were met by hundreds of other volunteers who inflicted significant casualties among the Regulars.

Patriots' Day is celebrated on the third Monday in April in Massachusetts with observances and re-enactments of these battles, considered the first of the Revolutionary War, and with the running of the Boston Marathon. The battles of Lexington and Concord have also been immortalized by Ralph Waldo Emerson in his 1867 poem "Concord Hymn," with the famous phrase, "the shot heard 'round the world," and by Henry Wadsworth Longfellow in his poem "The Midnight Ride of Paul Revere."

Within days, the Massachusetts Militia, a volunteer group of poorly trained but well-motivated citizens, surrounded the city of Boston to prevent the Royal Regulars from leaving. Munitions and supplies were sent from Georgia to aid the Massachusetts Militia and citizens of Boston.

Delegates to the Second Continental Congress met in Philadelphia in May. They established a Continental Army, made up of the militia units around Boston. George Washington was named commanding general, and he immediately began training volunteers. Even with the recent events of Lexington and Concord fresh in their minds, the Congress adopted and sent the Olive Branch Petition, a statement of loyalty to the king but disapproval of the actions of his ministers and Parliament which expressed hope for a reconciliation. King George III completely ignored it because by that time there was overt conflict between the Continental Army and the Royal Army and Navy.

The Second Continental Congress represented 13 of the 16 North American colonies. The colonies of East and West Florida and Quebec remained fiercely loyal to England and refused to attend. Citizens in the other colonies who refused to join in the rebellion fled south to the Florida colonies, the Bahamas and other Caribbean islands; the population of East Florida grew from 3,000 to 17,000 by the end of 1776. The Congress had no explicit legal authority since all the colonies were still under the direct governmental control of England, but it assumed all the functions of a national government, such as borrowing money, disbursing funds, and appointing military officers. Since the Congress had no authority or force to collect taxes and the colonies were slow to respond to any request for funds, it began issuing paper money called "continentals," which were designed and printed by Paul Revere. The Second Continental Congress met six times from 1775 to 1781.

Representatives from 10 of Georgia's 12 parishes met in Savannah in July at the colony's Second Provincial Congress and agreed to forbid trade with England and to establish an executive committee, the Council of Safety, which met when the Provincial Congress was not in session. The Council of Safety had the power to direct Georgia's military activities, undertake negotiations with Native Americans, issue money, and oversee the publication of a newspaper. Astonishingly still deferring to the paternalistic governor, the Council of Safety asked Wright to commission appointed officers into a militia rather than the Royal Regulars. When he refused, the Council acted immediately by commissioning them in the newly named Georgia Militia.

The delegates to the Second Continental Congress and most of the public were not yet convinced that armed rebellion was the way to handle their complaints. It wasn't until Thomas Paine, an Englishman who had been in the colonies little more than a year, published his pamphlet *Common Sense* in January 1776 that the spark which led to the public's conscience of revolt was kindled.

Paine, a jack-of-all-trades and not highly educated, turned himself into a philosopher and, because of his writings, became known as the Father of the American Revolution. *Common Sense* sold over 500,000 copies in North America and Europe and was the best-selling work of the 18th century. With an eloquence of style and persuasive argument, Paine convinced the delegates and the public to seek independence and not compromise. His strength lay in his ability to present complex ideas in clear and concise form. He wrote a series of inspirational essays from 1776 to 1783 called *The American Crisis* that begin with the famous words, "These are the times that try men's souls."

In February 1776, sensing his loss of political control and impending arrest, Georgia's Governor Wright and his family escaped to a Royal Navy ship holding offshore of Savannah at Tybee Island and returned to England. He died in 1785 at age 69 and is buried in Westminster Abbey in London. Wrightsborough, Ga., is named after him.

The Third Georgia Provincial Congress met in Augusta on May 1, 1776. Delegates elected a president and appointed George Walton and Button Gwinnett to join Lyman Hall at the Second Continental Congress.

During the early part of the summer, delegates in Philadelphia debated the decision to continue the war effort. In June a committee was formed to draft a document of independence, and Thomas Jefferson was chosen to write the paper. On July 4, 1776, the Congress formally endorsed the Declaration of Independence. Within a month, all 56 members of the Second Continental Congress, including delegates from Georgia, signed their names to the document.

Georgia's signers

Lyman Hall, the delegate from St. John's Parish, was elected as delegate to the Georgia House of Assembly in 1783 at the end of the war, and that legislature elected him governor. His suggestion that tracts of lands be set aside to establish educational academies in the future was instrumental in later chartering of the University of Georgia. After his year as governor, he served another year in the assembly, then a year as a judge before retiring to his plantations. Hall died in 1790 at age 66.

George Walton, the youngest signer of the Declaration of Independence, had an illustrious career. Born in Virginia, he moved to Savannah in 1769 to study law and developed a very successful practice. He was elected secretary of the Second Provincial Congress, and in December 1775 was chosen president of the Council of Safety. He served in the Continental Congress until October 1777, then returned to Georgia to defend its borders as a commissioned colonel of the First Regiment of the Georgia Militia. During the Battle of Savannah, he was injured and taken prisoner, then released in a prisoner exchange. After serving as governor of Georgia in 1779 for two months, Walton returned to the Congress in early 1780. At the end of the war in 1783, he came home to Georgia and was chosen as chief justice of the new state. He served as governor again in 1789 until a new government was begun under the new state constitution; then he was appointed a superior court judge. In 1795 he was appointed to fill the unexpired U.S. Senate seat of James Jackson until a successor was elected. Walton died at his home in Augusta in 1804 at age 68.

Button Gwinnett is one of the most famous signers of the Declaration of Independence. He was born in Gloucestershire, England, in 1732, and became a merchant in Bristol, England. He married in 1757 and sailed to Charleston five years later. He moved to the Savannah area in 1765 and started a plantation which became quite prosperous. Because of his close ties to England, Gwinnett had been a committed loyalist to the Crown, but perhaps due to his friendship with Hall, he became a strong advocate of colonial rights and played a pivotal role in Georgia politics. After signing the Declaration of Independence, Gwinnett returned to Georgia and was elected president of the Council of Safety in 1777.

His tragic story involves Lachlan McIntosh, Gwinnett's political rival during Georgia's Assembly years. McIntosh was born in Scotland, came to Georgia as a young child, and later had military training under Oglethorpe. He became involved in the revolutionary political movement and devoted himself to organizing the Georgia Militia for the defense of Savannah. During the early years of the Revolutionary War, he won prominence for his triumphs over the Royal Regulars coming up from Florida. It was because of his military experience that the Second Continental Congress promoted him, instead of Gwinnett, to the rank of Brigadier General in charge of the Georgia Militia.

Their paths collided when Gwinnett, as president of the Council of Safety, appointed himself to lead an assault on St. Augustine in East Florida to secure Georgia's southern border. McIntosh, hearing of the plan and believing it was politically motivated, tried to convince Gwinnett that this was ill-conceived and bound for failure. Gwinnett persisted and ordered McIntosh to accompany him, but it was too late. The expedition was a disaster with unnecessary loss of life and equipment. Georgia troops made a hasty retreat to Savannah.

During the next meeting of the Council of Safety, Gwinnett lost his bid for governor but was exonerated of any wrongdoing regarding the failed campaign. McIntosh called Gwinnett "a scoundrell and lying rascal" and openly criticized him for the failure of the Florida expedition. The next day Gwinnett challenged McIntosh to a duel which was fought on May 16, 1777.

Both McIntosh and Gwinnett were struck by the other's first ball: McIntosh in the leg, and Gwinnett in the hip, shattering the bone. While McIntosh recovered and went on to serve with distinction in the Continental Army and later retired to his plantation, loss of blood and infection resulted in Gwinnett's death three days later on May 19. He was the first of the Declaration of Independence signers to die. Because of his relative obscurity and his death at a young age, his signature is extremely rare and one of the most valuable American autographs.

Georgia's constitution

The Continental Congress recommended that all the newly independent states create their own government. Rather than debating this in the Provincial Congress, the Constitutional Convention met in Savannah in October 1776. Georgia was one of the first states to use the constitutional convention as a way to draft the document, with attendees breaking out into committees and eventually agreeing upon what was to be included. Although the colonists were used to the English form of government and the idea of a constitution was foreign to them, they nonetheless incorporated basic rights such as religious freedom, a representative government, freedom of the press, and trial by jury. Equally important, the attendees also included a method for amending their constitution. The final draft of the document was sent to the Provincial Congress in May 1777. It was adopted, and the Council of Safety was dissolved.

It was at this time that Georgia's 12 parishes were replaced by eight new counties, seven of which were named for prominent English politicians who supported the colonists' quest for autonomy. The eighth county, which included St. John's Parish, was named Liberty in honor of its early commitment to independence. Chatham County, where Savannah is located, is named after William Pitt, Earl of Chatham, the popular former prime minister of England who supported the colonists' position.

Georgia's first constitution remained in effect for 12 years until delegates met in Augusta to change the state constitution to conform to the U.S. Constitution.

Over the past 200 years Georgia has had 10 constitutions — more than any other state — with the most recent revision in 1983. The latest constitution incorporated provisions such as an equal protection clause, a requirement for uniform court rules, and nonpartisan election of judges. The Georgia Constitution also provides that all bills introduced in the legislature deal with only the matter named in the title of the bill, the so-called "single subject rule." This is unlike the U.S. Congress which allows "pork barrel" or "earmarks" which have nothing to do with the main subject of the bill but instead benefit the author's constituents for political gain.

The fight for independence

For a few years after the signing of the Declaration of Independence, little of military or political importance happened in Georgia. There were still many citizens loyal to the Crown in Georgia, and they slowly began an exodus to St. Augustine. Upon his return to England, Governor Wright argued successfully for a full-scale invasion of Georgia. By 1778 most of Georgia was back under English control, and Wright returned to Savannah to resume royal authority. Although Boston, New York, Philadelphia and Charleston were under royal control at one time or another, Georgia was the only one of the 13 revolutionary colonies to be under the Crown during most of the war. Much of the fighting in Georgia turned inland to Augusta, where the beleaguered Council of Safety government was trying to survive. Disorganized militia from Georgia and the Carolinas tried to resist, but the Royal Army prevailed and the city was under English control for a short time.

In 1777 Morocco became the first country to publicly recognize the United States, formalized by the Treaty of Peace and Friendship between the two nations, the longest unbroken treaty in U.S. history. The former U.S. embassy in Tangier, now the American Legation Museum, is the only building outside the United States designated a National Historic Landmark.

The signing of the Franco-American Alliance in May 1778 not only provided much-needed money and troops for the colonies but also served as a tremendous psychological boost. Historians argue about the reasons the French chose to align with the struggling revolution, but most agree that France sought revenge for the loss of all its territory in the New World at the end of the war 15 years earlier and hoped to get Canada and Louisiana back if the colonies were successful.

Spain also provided money and supplies to the colonies to protect its presence in South America and to regain Florida, lost at the end of the French and Indian War.

The Netherlands had been an English ally for over a hundred years, but the spark of colonial independence brought great sympathy from the Dutch public. The Netherlands had been trading with the colonies for almost 50 years and refused England's request to boycott the rebellious colonies. It was one of the first countries to recognize the United States of America by lending Congress two million American dollars.

By 1780, after four difficult years of war, the colonies' chances for independence seemed slim. The strength of the Royal Army and Navy and their seemingly unlimited resources was overwhelming. The weak Continental Congress, exhaustion of the troops, and lack of funds and supplies all appeared to be insurmountable problems. However that changed in the blink of an eye.

The Spanish capture of the English fort at Pensacola, East Florida, in May 1781 provided a huge boost for the Continentals. The Royal Regulars, retreating from Florida, joined General Charles Cornwallis' troops in Yorktown, Va., where the exhausted troops were to be picked up and returned to England by the Royal Navy standing offshore. Learning of this, Washington sent a fleet of 48 French ships to Chesapeake Bay which defeated the Royal Navy in a fierce battle — the only major naval defeat suffered by the Royal Navy in 200 years. With the French Navy in control of the bay's entrance, 20,000 French and Continental forces marched down from New York and surrounded Cornwallis and his Regulars. After an artillery duel of only three weeks and with minimum losses on both sides, Cornwallis surrendered to Washington on Oct. 19, 1781. Over 8,000 English troops — a quarter of all English forces — became prisoners of war. This humiliating defeat of the powerful royal military force was the last major battle of the Revolutionary War.

King George III wanted to continue the fighting after the Cornwallis defeat, but when his prime minister Lord North resigned in protest, Parliament and public sentiment insisted that the peace process begin. The United States sent John Adams, Benjamin Franklin and John Jay to Paris to negotiate a treaty.

James Oglethorpe, the founder of Georgia who had returned to England four decades earlier, was also a supporter of the colonists and in 1785 paid his respects to Adams, the first U.S. minister plenipotentiary to the Court of St. James's (ambassador to Britain), shortly after his arrival in London.

Though the war was essentially over, skirmishes continued throughout the colonies. News traveled slowly, and even as the negotiators worked in Paris to resolve the issues, fighting occurred well into 1782. Almost 10,000 loyalists and slaves left Savannah after the defeat at Yorktown, and Wright with his family and government left Georgia in July 1782. The English control would soon come to an end, and it would be up to the newborn states to live up to their own expectations.

A new nation emerges

On Sept. 3, 1783, the Treaty of Paris formally ended the Revolutionary War. Among other things, the treaty provided for England's recognition of the 13 colonies to be free and independent states, specified that the new United States have fishing rights off the coast of Newfoundland, allowed free navigation on the Mississippi River for both England and the United States, and required the return of confiscated property to those colonists who had remained loyal to the Crown.

Determining the borders for the new nation, however, was not easy because very few maps were available. The well-known Mitchell Map, named for its creator, was used during the treaty negotiations. John Mitchell was born in Virginia in 1711, was sent to Scotland for his education, and graduated from the University of Edinburgh. It is unclear whether he ever received a medical degree, but he did return to Virginia to practice medicine. Because of ill health, he moved to England in 1746 where he gained prominence for his interest in botany and cartography. He was commissioned by the Board of Trade and Plantations in 1749 to make a map showing England's claims in North America. Compiling data from other maps and reports, he produced a remarkably accurate "Map of the British and French Dominions in North America" first published in 1755. (Mitchell died in 1768, but several editions of his map with mostly minor edits and changes were made by various publishing companies during the late 18th century. All are known as the Mitchell Map.)

Article 2 of the 1783 Treaty of Paris declares that the boundaries

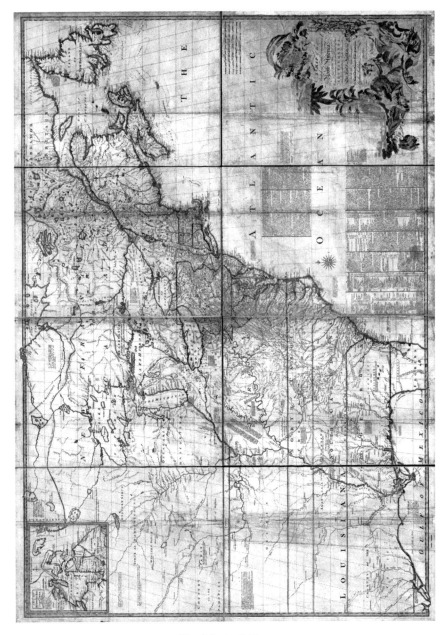

The Mitchell Map

Library of Congress, Geography and Map Division

of the United States are

... from the northwest angle of Nova Scotia, viz., that angle which is formed by a line drawn due north from the source of St. Croix River to the highlands; along the said highlands which divide those rivers that empty themselves into the river St. Lawrence, from those which fall into the Atlantic Ocean, to the northwesternmost head of Connecticut River; thence down along the middle of that river to the forty-fifth degree of north latitude; from thence by a line due west on said latitude until it strikes the river Iroquois or Cataraquy; thence along the middle of said river into Lake Ontario; through the middle of said lake until it strikes the communication by water between that lake and Lake Erie; thence along the middle of said communication into Lake Erie, through the middle of said lake until it arrives at the water communication between that lake and Lake Huron; thence along the middle of said water communication into Lake Huron, thence through the middle of said lake to the water communication between that lake and Lake Superior; thence through Lake Superior northward of the Isles Royal and Phelipeaux to the Long Lake; thence through the middle of said Long Lake and the water communication between it and the Lake of the Woods, to the said Lake of the Woods; thence through the said lake to the most northwesternmost point thereof, and from thence on a due west course to the river Mississippi; thence by a line to be drawn along the middle of the said river Mississippi until it shall intersect the northernmost part of the thirty-first degree of north latitude, South, by a line to be drawn due east from the determination of the line last mentioned in the latitude of thirty-one degrees of the equator, to the middle of the river Apalachicola or Catahouche; thence along the middle thereof to its junction with the Flint River, thence straight to the head of Saint Mary's River; and thence down along the middle of Saint Mary's River to the Atlantic Ocean; east, by a line to be drawn along the middle of the river Saint Croix, from its mouth in the Bay of Fundy to its source, and from its source directly north to the aforesaid highlands which divide the rivers that fall into the Atlantic Ocean from those which fall into the river Saint Lawrence; comprehending all islands within twenty leagues of any part of the shores of the United States, and lying between lines to be drawn due east from the points where the aforesaid boundaries between Nova Scotia on the one part and East Florida on the other shall, respectively, touch the Bay of Fundy and the Atlantic Ocean, excepting such islands as now are or heretofore have been within the limits of the said province of Nova Scotia.

These boundaries were essentially that of the English territory in the prior Treaty of Paris that had ended the French and Indian War. The notable exception was that both of the Florida colonies were returned to the Spanish as reward for their assistance. At this time, Spain also controlled the entire Louisiana Territory, which included New Orleans and the mouth of the Mississippi River. The Spanish territory would be a problem for the United States for several decades.

After the Revolutionary War, 1783

Western expansion

With the signing and subsequent ratification of the 1783 Treaty of Paris by all the involved governments, the Revolutionary War was officially over, but Georgia was left in a confused and deplorable state. The countryside had been severely ravished, Georgia was loaded with debt, collecting taxes was difficult, and politicians struggled to set up a workable government. Ordinary citizens were now focused on making a living and dealing with the Native Americans' demands and with their slaves who were looking for freedom. In addition, loyalists to the Crown were coming back from Florida to reclaim their land. Settlers from other colonies had been coming into Georgia and inhabiting land along the Atlantic coast between the Savannah and Altamaha rivers for the past 50 years, but now that the English were gone, the interior lands were being explored and settled.

Georgia's boundaries had changed several times from the original charter in 1732, yet none of these boundaries had been surveyed. Indeed, most of what the colonists thought was Georgia's land was occupied by native tribes and had rarely been seen by any white man. The lure for profits from buying and selling the western lands brought new investors into the state.

In 1784 the Georgia General Assembly granted speculators a tract of land, thought to be in Georgia, at Muscle Shoals on the Tennessee River. The investors named the area Houston County, and commissioners were appointed to survey the area and establish a county government. Some of the commissioners and a few of the land speculators traveled to the region; a land sales office was set up, but no surveys were ever done. For reasons not completely understood but probably related to trouble with the tribes who lived there as well as claims by North and South Carolina that the land was within their boundaries, the whole project was abandoned. In 1786 the General Assembly directed the commissioners to return to Augusta, Georgia's capital.

Another attempt to sell and settle lands in the western part of Georgia occurred in February 1785. The General Assembly passed an act, signed by Governor George Matthews, authorizing justices of the peace and probates to be sent to an area on the Mississippi River near present-day Natchez to establish Bourbon County, named in honor of the Bourbon monarch of France who had been a valuable ally to the colonists' independence. Soon after, however, Georgia repealed any legislation regarding the county at the strong urging of the U.S. Congress because of significant problems with the Native Americans and the Spanish.

The Beaufort Convention

For decades, Georgia and South Carolina had argued over who had navigational rights on the Savannah River and the exact location of its headwaters. The original language in the Trustee's charter was vague as to where the Savannah River started and, as settlers were moving into the area, an exact boundary line was needed. Both states appealed to the U.S. Congress for assistance in deciding the dispute; after months of delay, they were instructed to meet with each other to resolve the conflict. Lachlan McIntosh, who had fatally wounded Button Gwinnett in their duel, was appointed a member of the Georgia delegation.

In April 1787 commissioners from the two states met in Beaufort, S.C. Within a few days they agreed on what they believed to be more concise language regarding their boundary. Known as the Beaufort Convention or the Treaty of Beaufort, this definition was:

> The most northern branch or stream of the River Savannah from the sea or mouth of such stream to the fork or confluence of the Rivers then called Tugaloo and Keowee; and from thence the most northern branch or stream of said River Tugaloo, till it intersects the northern boundary line of South Carolina, if the said branch or stream of Tugaloo extends so far north, reserving all the islands in the said Rivers Savannah and Tugaloo, to Georgia; but if the head, spring, or source of any branch or stream of the said river Tugaloo does not extend to the north boundary line of South Carolina, then a west course to the Mississippi, to be drawn from the head, spring, or source of the said branch or stream of Tugaloo river, which extends to the highest northern latitude, shall forever thereafter form the separation, limit, and boundary between the States of South Carolina and Georgia.

Since the Carolinas had agreed as early as 1735 that the boundary between North and South Carolina would be the 35th parallel north, the reference in the Beaufort Convention to the "northern boundary line of South Carolina" could be inferred to mean the boundary line between Georgia and South Carolina was also the

35th parallel.

Although the Chattooga River is not mentioned in the treaty, it is an important factor in determining the starting point of the Georgia/South Carolina boundary. The Chattooga begins near Cashiers, N.C., and flows south, crossing 35° north latitude to join with the Tallulah River to form the Tugaloo River. In the 18th and 19th centuries, the Tugaloo continued to flow southeasterly to join the Keowee River, the confluence of which formed the Savannah River. (That changed in 1923 when Georgia Power's hydroelectric dam created Lake Tugaloo by damming the Tugaloo River so that the confluence of the Chattooga and Tallulah rivers are now under the middle of the lake. In 1963, Lake Hartwell was formed by damming the Savannah River so that the meeting of the Tugaloo and Keowee rivers are now under Lake Hartwell.)

Georgia lost thousands of acres of land by agreeing to accept the Tugaloo River as the most northern source of the Savannah River, since the Keowee River rises far more north of the 35th parallel than the Tugaloo. By choosing the Tugaloo River, Georgia lost a

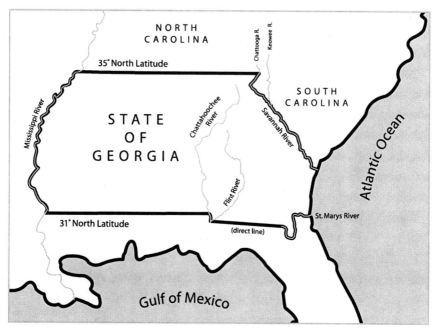

After the Beaufort Convention, 1787

large triangle of land that is now South Carolina.

As part of the 1787 Beaufort Convention, South Carolina also ceded all western lands between the North Carolina/South Carolina boundary line and the southern branch of the Tugaloo River to the United States, who then ceded this land to Georgia in 1802. In reality, South Carolina ceded nothing to the United States because the Tugaloo River system rises north of the 35th parallel and is in North Carolina. This caused confusion for settlers who came to that area because they did not know in which state they lived.

Establishing a government

The six-year period after the Treaty of Paris in 1783 was a time of disarray for the colonies. The Continental Congress was officially disbanded and renamed the Confederation Congress, but this assembly had no legitimate authority to govern the colonies. With England no longer in control, the 13 new states found themselves quarreling over such important issues as boundaries, militia, taxes, legislation and administration. In 1787 with only Rhode Island refusing to attend, 12 of the former colonies met in Philadelphia and voted to create an entirely new form of a federal government. On Sept. 17, 1787, the delegates ratified the final draft of a Constitution and sent it to all the states for approval.

Nine of the 13 states were needed to endorse the new document, but not all of them embraced the idea of a federal government. Over an eight-month period, a series of editorials were written by James Madison, John Jay and Alexander Hamilton to gain support for the new Constitution and the idea of united rather than individual states. "The Federalist Papers" appeared in the *New York Independent Journal* and were widely circulated through the states.

Georgia was the fourth state to ratify the new Constitution in January 1788, and by June, with New Hampshire's ratification, the Constitution became valid. This initiated elections of the legislative, judicial and executive branches of the United States. All 13 states eventually ratified the Constitution, although Rhode Island and North Carolina did so after the presidential inauguration.

On April 21, 1789, in New York City, John Adams was sworn in as vice president. Nine days later, after an arduous journey from Virginia, George Washington was sworn in as the first president of the United States of America. The time had come for a new nation to face the enormous challenges of dealing with its own intrinsic issues as well as becoming part of the world community.

The birth date of Georgia

Many historians have written about what date should be celebrated as the state of Georgia's birth date. The first mention of the name Georgia found in the *Journal of the Transactions of the Trustees* was Feb. 13, 1730. The date when King George II signed the Trustees' charter was April 21, 1732. Oglethorpe and the first colonists came to Yamacraw Bluff to begin clearing land for the site of Savannah on Feb. 1, 1733. It could be considered July 4, 1776, when independence from England was declared, or Jan. 2, 1788, when Georgia ratified the U.S. Constitution and was recognized as the fourth state of the young nation.

To make things more complicated, however, the difference between the Julian and Gregorian calendars must be taken into account. In 45 B.C., Julius Caesar introduced a calendar of 364 days which was observed until 1582, when Pope Gregory instituted a new calendar after meeting with his astronomers and mathematicians. For years the Italian scholarly community had noted that the actual time for Earth's rotation around the sun was not 364 days but rather a little more than 365 days. Over the centuries since Caesar's calendar, this additional time added up to 10 extra days. Pope Gregory's advisors corrected the problem by omitting the 10 days from the current year and adding an extra day to the calendar every four years to make up for the additional time. All the Catholic countries immediately obeyed the papal decree, but it was not until 1752 that the day after Sept. 2 became Sept. 13 in England and the colonies.

Whether to use the Julian or Gregorian calendar was finally solved in 1909, when the Georgia General Assembly provided that

> ... the twelfth day of February in each year shall be observed in the public schools of this State, under the name "Georgia Day," as the anniversary of the landing of the first colonists in Georgia under

Oglethorpe, and it shall be the duty of the State School Commissioner through the County School Commissioners, annually to cause the teachers of the schools under their supervision to conduct on that day exercises in which the pupils shall take part, consisting of written compositions, readings, recitations, addresses, or other exercises, relating to this state and its history and to the lives of distinguished Georgians. When said day falls on Sunday, it shall be observed on the following Monday.

Important Dates III

1763	Proclamation Act
	Colonies of East and West Florida and Quebec established
1764	Sugar Act
1765	Stamp Act
	All colonies have Committees of Correspondence
1766	Declaratory Act
1767	Townshend Act
1773	Tea Act
1774	First Continental Congress
1775	Parliament declares American colonies in rebellion
	First meeting of Second Continental Congress
	Battles of Lexington and Concord
1776	*Common Sense* published
	James Wright, last royal governor of Georgia, returns to England
	Declaration of Independence written and signed; Colonies formally declare war on England
1777	First Georgia Constitution adopted by Provincial Congress
	First Georgia General Assembly
1778	France declares war on England
1779	Spain declares war on England
1781	Continental Congress renamed Confederation Congress
	Articles of Confederation adopted
1783	Treaty of Paris ends Revolutionary War
1785	University of Georgia, first state-chartered university in the United States, incorporated
1787	South Carolina cedes land to the United States

	U.S. Constitution ratified at Confederation Congress
	Beaufort Convention between Georgia and South Carolina
1788	Georgia ratifies U.S. Constitution
1789	Second Georgia Constitution adopted
	Inauguration of President George Washington

IV
Defining Georgia

One could almost hear a collective sigh of relief after the new government of the United States was brought to life with a new constitution and a new president, but the growing pains of a young nation were just beginning. In 1789, 90% of the population of the United States lived on the Atlantic coastline, and the fertile land west of the Allegheny Mountains was beckoning. Settlers and investors were pouring westward. Land acquisition was on everyone's mind.

It was probably no coincidence that claiming more land for the United States was important to George Washington and Thomas Jefferson; both were trained surveyors and owners of large land tracts. The Confederation Congress had passed Land Acts of 1784, 1785 and 1787 reflecting the importance the new nation gave to acquiring land west of the Alleghenies. Although many settlers negotiated to buy their land, when it came to a treaty between Native Americans and the U.S. government, more often than not, the terms would be violated by the government and the people forcibly removed from their territory. By 1835 over 400 separate land cession treaties were negotiated between the United States and the tribal nations. The United States wanted all the land from the Atlantic coast to the Mississippi River.

Surveying the land played an integral role as this territory was divided. Advances in cartography, astronomy and mathematics meant it was now possible to define and locate fixed points other than natural landmarks, and a variety of events, including disputes, treaties and agreements, as well as the creation of new states necessitated clear border lines. Within three decades, all of Georgia's boundaries were measured and marked. By the late 1830s, the now-familiar and final shape of the state had emerged.

Surveying

It is difficult to appreciate the difficulties surveyors experienced. They walked for miles through rough terrain, dense forests and swamps, blazing a trail as they went, and dealt with tribes who did not welcome their presence. The surveyors had to carry or shoot their own food, find potable water, live in tents under all types of weather extremes, and endure terrible sanitary and hygienic conditions. Their instruments were relatively crude and simple, as was their knowledge of astronomy and mathematics.

A survey team consisted of axmen, cooks, common laborers, armed guards, and perhaps most importantly the chain carriers who measured distances along the surveyed line. Known as the Gunter's chain, the survey chain was conceived and fabricated in the early 1600s by Edmund Gunter, a mathematician and professor of astronomy at Gresham College in London. The chain is 66 feet long and divided into 100 equal links for measuring intermediate lengths. The mathematics using the chain are extremely easy: 80 chains are one mile (80 × 66 feet = 5,280 feet) and 10 square chains are one acre (10 × 66 feet × 66 feet = 43,560 square feet). The Gunter chain was used by surveyors until the early 20th century.

Latitude and longitude

Surveyors had to be knowledgeable about latitude and longitude. Latitude can be determined by locating the North Star and measuring the angle between the North Star and the horizon using instruments such as a quadrant or sextant. Maps are drawn such that latitude lines circle the globe running east and west; these horizontal lines are also called parallels because latitude lines are parallel to each other. The equator is the 0° line of latitude; the North Pole is at the 90th parallel of north latitude, and the South Pole is located at the 90th parallel south. Each degree of latitude is further subdivided into minutes and seconds so that the designation of a point on Earth, such as the city of Atlanta, would be latitude 33 degrees, 46 minutes, 47 seconds north (33° 46' 47" N).

Because the distance between the parallels is constant, they can also be used to measure distance north and south. Each degree of latitude is 60 nautical miles, so the distance from the equator to Atlanta is about 1,980 nautical miles (33 degrees × 60 nautical miles = 1,980 nautical miles).

Measuring latitude was relatively easy, but knowing where along a parallel of latitude one was located was much more difficult, and virtually impossible at sea. When Christopher Columbus left Spain to find the direct route to Asia by sailing to the west, he was able to determine his latitude so that he could keep his course on a straight line, but he had no idea how far he had traveled from Spain. Unlocking the secret of longitude took hundreds of years.

Picture a 360-degree circle around the Earth at any latitude, divided into 24 equal segments, each representing one hour in a day. A segment is then 15 degrees (360 degrees ÷ 24 segments = 15 degrees/segment), and therefore each degree is 4 minutes of time (60 minutes ÷ 15 degrees = 4 minutes/degree). While latitude has a fixed starting point at the equator, a series of 24 longitude segments could begin at any point on the globe.

The problem of determining longitude at sea was solved by the English clockmaker John Harrison. He spent his life trying to win a

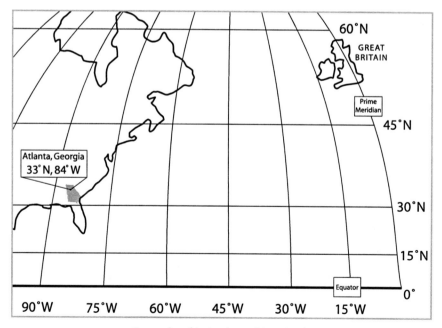

Example of latitude and longitude

prize offered by Parliament and in 1761 invented an extremely reliable clock which sailors could set with the time in London and then take aboard ship to compare with the local time, determined by the position of the sun. Nineteenth-century navigators converted the difference in hours and minutes into degrees to give them their longitude. If the London clock said the time was noon and the local time was 6:00 a.m., six hours earlier, the longitude location was 90 degrees east of London since every hour is 15 degrees (6 hours × 15 degrees/hour = 90 degrees).

A given longitude along a latitude parallel provides a specific position that can be located consistently. Our modern time zones are also based on these longitude lines, or meridians.

When the longitude-time connection was finally established, each country began to measure longitude starting from their own zero meridian. In 1791 Secretary of State Thomas Jefferson wrote a letter to the well-known surveyor Andrew Ellicott, who had been hired to survey and plat the Federal Territory (now known as the District of Columbia), to determine a 0° longitude within the Territory to serve as the nation's prime meridian. The location of the United States' first 0° longitude was in the center of the site proposed for the new Congress House. Ellicott's original plat of the Territory shows the longitude "noted to be 0° 0' and situated 77° 00' 33.5333" west of Greenwich, England." Jefferson believed that using the Federal Territory 0° longitude as the prime meridian would further separate the young nation from any ties to England. However, the main work of surveying the Territory and designing the city of Washington overshadowed the location of the prime meridian, and it wasn't until 1804 when Thomas Jefferson was president that a monument known as the Jefferson Stone was erected. It never served as the nation's prime meridian, but it was used by local surveyors as a point of beginning.

In 1884 an International Meridian Conference convened at Washington, D.C., attended by delegates from 21 nations. To facilitate international commerce and travel, they agreed that a common starting point of longitude was necessary to provide a standard of worldwide time, and they set the location of the global prime meridian at the Royal Greenwich Observatory just outside of London. Surveyors, cartographers, navigators and travelers all now use the same longitude and time system. Millions of tourists visit Greenwich annually and have their picture taken astride the prime meridian.

The Pinckney Treaty

The most serious and ongoing dispute over land was between Spain and the United States. For almost 15 years since the end of the Revolutionary War, Spain remained angry that the boundary of West Florida was not the 1767 boundary of 32° 28' north at the Yazoo River but instead the original 31° north latitude almost a hundred miles south. The United States' position was that giving the Florida colonies to Spain was a generous gift of appreciation and that the boundaries of the gift would be decided by the giver. In retaliation, Spain encouraged its citizens in the area to trade with the local tribes and sell them weapons and ammunition to use in resisting the arrival of new settlers coming from Georgia and the Carolinas. Spain also refused to allow the United States navigation on the lower Mississippi River and access to the Spanish port at New Orleans.

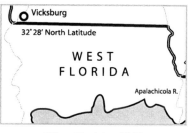

West Florida, 1767

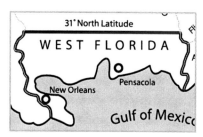

Original West Florida, 1763

After years of attempted and failed negotiations regarding the border, Spain noticed that the relationship between the United States and England was becoming stronger, and, with a growing concern that the two nations would attempt to take over Spanish territories west of the Mississippi, took an interest in having a positive relationship with the United States. In 1795 the two finally signed a treaty regarding the West Florida border. Known as the Pinckney Treaty, named after South Carolinian Thomas Pinckney, the envoy extraordinary to Spain, it was an extremely important step for the young nation. Spain agreed to accept the 31° north latitude line as its north border. Both countries agreed that each would have free navigation on the Mississippi River and that the United States would have complete use of the port of New Orleans. Both countries also set the border between the United States and Spanish Louisiana as the middle of the Mississippi River and agreed to appoint commissioners to jointly survey the line within six months of signing the treaty. In addition, both countries agreed not to incite the Native Americans to warfare.

Andrew Ellicott

Almost immediately after the treaty was ratified by the Congress, President George Washington commissioned Andrew Ellicott to survey the new border between Spain and the United States. Ellicott was already recognized as a competent surveyor. He had surveyed the 10-square-mile boundary for the Federal Territory and completed measuring the line between Maryland and Pennsylvania, which Mason and Dixon had started but abandoned because of problems with the local tribes. A third-generation American who had been a major in the Revolutionary War, Ellicott was a self-taught mathematician, astronomer and clockmaker — skills which aided him as a surveyor. He made most of his personal astronomical and surveying instruments himself. The four years he spent surveying the United States/Spain border, part of which would later be the Georgia/Florida boundary, is one of the most amazing yet largely untold stories in American history.

Ellicott had been living in Philadelphia for six years with his wife and nine children when he received the letter from Washington to survey the new border. In September 1796 he and his teenage son, Andrew Jr., traveled by horseback the 350 miles to Pittsburgh, where he assembled a team of cooks, axmen, chain carriers and a military escort of 30 men and began the trip down the Ohio and Mississippi rivers on four sailing vessels. They arrived at Natchez, an important Spanish diplomatic post on the Mississippi River, in February 1797 where they met the Spanish survey team.

Ellicott spent an entire year in Natchez dealing with the Spanish governor and his staff, who pretended that they were unaware of the 1795 treaty and refused to withdraw from the Spanish capital or to allow Ellicott to begin the survey. Finally, on April 8, 1798, the Spanish governor and his troops departed from Natchez, leaving Ellicott and the Spanish surveying team to commence the measurement of the border. Starting some 30 miles south of Natchez, at 31° north latitude, both teams set up their camps and began their survey. The work was very difficult and involved surveying, making astronomical observations to determine latitude and longitude, marking a line through the woods and swamps, and dealing with hostile native populations.

Seven months later, the team arrived at the Pearl River, a mere 70 miles east of the Mississippi River. On instructions received by letter from Secretary of State Thomas Jefferson to map out the rivers draining into the Gulf of Mexico, Ellicott went to New Orleans. He supervised the construction of a 30-ton boat, which he named *Sally* in honor of his wife. He also bought tents, horses, clothing and other supplies for the teams. Since he did not know how to sail, Ellicott hired two deserters from an English privateer as crew to help him sail the *Sally* from New Orleans to Mobile Bay and then up the Mobile River to where the teams were now located. After satisfying himself that the team's surveyed line was correctly placed on the 31st parallel, he marked the location with a monument.

The 200-year-old monument just north of Mobile, Ala., known as the Ellicott Stone, is three feet high and engraved on the north side with "U.S. Lat. 31 1799" and on the south side with "Dominos de S.M.C. CAROLUS IV Lat 31 1799" (Dominion of his Majesty King Charles IV). This is the only known stone monument set by Ellicott during the entire time he surveyed this boundary. A hundred years later, surveyors used the Ellicott Stone as the initial point to start the surveying of all public lands in the southern region of Alabama and Mississippi. The Ellicott Stone was placed on the National Register of Historic Places in 1973.

Ellicott and his team continued surveying and marking the 31st parallel going eastward until they came to the Chattahoochee River and then, according to the Pinckney Treaty, went down the Chattahoochee until it joined the Flint River, always making astronomical and surveying observations. However, the ongoing problem with angry tribes caused Ellicott great concern. He divided up the surveying team such that one group of surveyors, with his son Andrew Jr. and armed guards, would take the horses and continue marking the line eastward towards the source of the St. Marys River, while Ellicott and his team would sail around Florida and find the mouth of the St. Marys as it flows into the Atlantic Ocean.

After sailing down the Apalachicola River into the Gulf of Mexico, Ellicott sailed east to St. Marks, Fla., where he obtained more supplies for the trip. Accompanying him was a military escort plus 20 persons, only five of whom had ever been to sea before. Of those five, two were "sailors and totally illiterate."

Ellicott left St. Marks on Oct. 16, 1799, and for the next six weeks sailed down the western coast of the Florida peninsula to Cape Sable then turned east around Key Biscayne and north up the eastern coast. He had no information about the St. Marys River, but he did have, as he writes in his journal, "a small mutilated chart and St. Marys appeared to be about the 31st latitude north." On Dec. 3, 1799, Ellicott turned the *Sally* into a channel of the Atlantic Ocean and landed at St. Simons Island. After spending several days exploring the ruins of Fort Frederica and replenishing supplies to finish the survey, Ellicott sailed south to the village of St. Marys where he found his son and the other half of the survey team, who had walked from the Chattahoochee and Flint River junction through the Okefenokee swamps, waiting to meet him.

On Jan. 23, 1800, after making additional astronomical observations, Ellicott and the entire team left the village and proceeded up the St. Marys River as far as they could on the schooner, then packed their instruments and supplies into canoes and rowed further into the swamp. Ellicott found that it was "impossible with our few remaining broken down pack horses to convey our apparatus by land to the source of the river St. Marys as it is formed by the water draining out of the Okefenokee swamp along several marshes or small swamps which join into one and form or constitute the main branch or body of the river. The swamps are impenetrable without immense labor and expense." Using mathematical triangulation, Ellicott chose a point which he believed was the source of the St. Marys River, but he knew that his location was just an educated guess and that later measurements, with improved instruments, might prove him wrong.

With the final measurement being made, Ellicott wrote in his journal, "The 26th. Capt. Minor his catholic Majesty's commissioner and myself, with a party of labourers went to the Swamp and the day following had a mound of earth thrown up on the west side of the main outlet and as near to the edge of the swamp as we could advance on account of the water and the alligators." Ellicott marked the location of the source of the St. Marys River at "30 degrees, 34 minutes 48 seconds latitude north, and 5 hours, 29 minutes and 9 seconds west of Greenwich." On the same day, Ellicott and the team returned to St. Marys.

After arriving in the village, the surveying team, the *Sally* and the military escort were sent back to the newly established U.S. fort on the Mobile River. Ellicott spent several weeks on Cumberland Island preparing reports for both Spain and the United States. He had made over 400 observations over the preceding two years, and his obsession with accuracy demanded that he precisely prepare his transcripts. Finally he and his son arranged to be taken to Savannah to board a commercial sailing vessel for Philadelphia where they arrived on May 18, 1800, three years and seven months after they had left.

Within a few weeks of being home, Ellicott requested payment for his services and presented his report to President John Adams. He was offered an appointment as surveyor general of the United States but refused the position, complaining in a letter to Adams that because his pay had been withheld he had to sell his valuable library and dispose of one of his surveying instruments to "procure money for market." His friend Thomas Jefferson, now vice president, finally became involved and returned all of Ellicott's records to him so that he could publish his report. It is unclear whether he was ever paid for his monumental service to the United States.

In 1803 Ellicott published the 450-page record of his journey entitled *The Journal of Andrew Ellicott, for Determining the Boundary between the United States and the Possessions of His Catholic Majesty In America containing occasional remarks on the Situation, Soil, Rivers, Natural Productions, And Diseases of the Different Countries on the Ohio, Mississippi, and Gulf of Mexico with Six Maps to which is added An Appendix Containing all the Astronomical Observations Made Use Of for Determining the Boundary*. His daily diary along with 150 pages of scientific data describe not only the hardships endured on his journey but also reveal his wide range of knowledge in areas such as politics, geography, botany, cartography, astronomy, medicine, surveying, mathematics and meteorology. Most Americans have never heard of Andrew Ellicott, but his contributions to astronomy and surveying make him a giant in U.S. history.

The Yazoo Land Fraud

The most infamous land scandal in the United States occurred in Georgia in 1795. Georgia's legislators passed a law selling 40 million acres of land, located where the Yazoo River entered the Mississippi River, for ridiculously low prices to investment companies whose major stockholders were Georgia legislators and officials. This land was then resold to private individuals at much higher prices. Public outrage grew because of the huge profits being made by the officials, and soon the involved legislators, including the governor, were removed from office.

Within a year, the Georgia legislature repealed the act and nullified all the land contracts which had been made, but the private owners refused to give up their land claiming that their purchase contracts were valid. Georgia refused to recognize the ownership of the lands, and the U.S. Supreme Court in 1810 ruled that the sales were in fact binding contracts and could not be retroactively invalidated.

Tennessee

In 1796 the U.S. Congress admitted Tennessee as the 16th state with the following language:

> The whole of the territory ceded to the United States by the State of North Carolina shall be one State and the same is herby declared to be one of the United States of America and on equal footing with the original states in all respects whatever, by the name and title of the State of Tennessee.

The state's constitution described the southern boundary as "that boundary which is described in the Act of Cession of North Carolina to the United States."

Both Tennessee and North Carolina abutted Georgia's northern boundary, and although there was no mention of the exact location of Tennessee's boundary in either the United States Acts or Tennessee's Constitution, North Carolina's southern boundary had been the 35° north latitude for 60 years. Georgia would depend on this fact to believe its northern boundary with both states was the 35th parallel.

Mississippi

As Ellicott was surveying the United States/Spain border, the United States created the Mississippi Territory in April 1798 by buying — or blatantly taking — land on which the Chickasaw and Choctaw nations lived. The new territory's southern boundary was the familiar 31st parallel north, stretching from the Mississippi to the Chattahoochee River, and the northern boundary was 32° 28' north latitude, the old northern boundary of West Florida, although this land also belonged to Georgia.

In May 1798 the Georgia Constitution was amended for the third time to include the official description of the state's boundaries as:

> The limits, boundaries, jurisdictions, and authority of the State of Georgia do, and did, and of right ought to, extend from the sea or mouth of the river Savannah, along the northern branch or stream thereof, to the fork or confluence of the rivers now called Tugalo and Keowee, and from thence along the most northern branch or stream of the said river Tugalo, till it intersect the northern boundary line of South Carolina, if the said branch or stream of Tugalo extends so far north, reserving all the islands in the said rivers Savannah and Tugalo to Georgia; but, if the head spring or source of any branch or stream

of the said river Tugalo does not extend to the north boundary line of South Carolina, then a west line to the Mississippi, to be drawn from the head spring or source of the said branch or stream of Tugalo River, which extends to the highest northern latitude; thence, down the middle of the said river Mississippi, until it shall intersect the northernmost part of the thirty-first degree of north latitude; south, by a line drawn due east from the termination of the line last mentioned, in the latitude of thirty-one degrees north of the equator, to the middle of the river Apalachicola, or Chatahoochee; thence, along the middle thereof, to its junction with Flint River; thence straight to the head of Saint Mary's River; and thence, along the middle of Saint Mary's River, to the Atlantic Ocean, and from thence to the mouth or inlet of Savannah River, the place of beginning; including and comprehending all the lands and waters within the said limits, boundaries, and jurisdictional rights; and also all the islands within twenty leagues of the seacoast.

Georgia ignored the new boundaries of the Mississippi Territory carved out of its land just one month earlier.

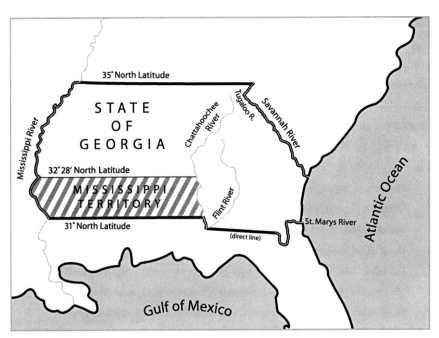

After Georgia Constitution amended, 1798

Georgia land cession

For years, Georgia had been interested in selling its western lands either to private investors or to the United States, but negotiations had always been unsuccessful. In the 1802 Articles of Agreement and Cession, Georgia agreed to cede all its western lands, including the Mississippi Territory, to the United States for $1,250,000, and the United States agreed to bear all expenses needed to settle Native American or Yazoo lands claims within the ceded area. Georgia was

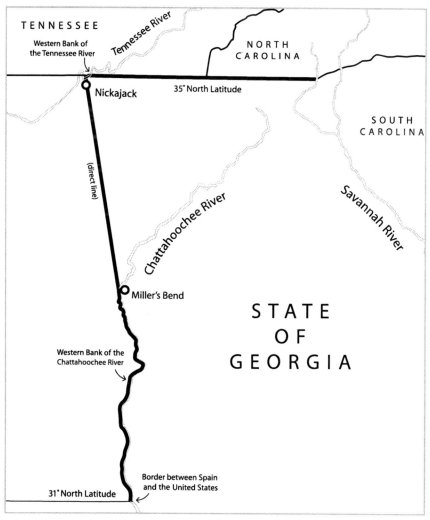

Northern and western boundaries
after Articles of Agreement and Cession, 1802

the last of the 13 original colony-states to come to an agreement with the United States regarding land cessions and the only one to receive money for doing so.

As a result, for the first time since 1732, the northern and the western boundaries of Georgia changed. They were now

> ... beginning on the Western bank of the Chatahoochie River where the same crosses the boundary line between the United States and Spain; running thence up the said River Chatahoochie, and along the Western bank thereof to the great bend thereof, next above the place where a certain creek or river called the "Uchee" (being the first considerable stream on the Western side, above the Cussetas and Coweta towns), empties into the Chatahoochie River, thence in a direct line to Nickajack, on the Tennessee River; thence crossing the said last-mentioned river, and thence running up the said Tennessee River and along the Western Bank thereof to the Southern Boundary line of the State of Tennessee.

Note that the western bank of the Chattahoochee River, not the middle of the river, was used as the western boundary.

The "nonexistent land"

When South Carolina ceded to the United States a 50-mile-wide strip of land believed to be between the northern border of Georgia and the southern border of North Carolina as part of the Beaufort Convention in 1787, numerous settlers headed west of the Allegheny Mountains around the French Broad River and near present-day Cashiers, N.C. For 15 years, this area was populated by people who had no government and had been given or purchased land grants from either Georgia or the Carolinas. Although the United States agreed to give Georgia this land in 1802, North Carolina also claimed it. Because the area was never surveyed and marked, this strip of "nonexistent land" was the focus of dispute and bloodshed.

Many of the settlers there wanted to be part of Georgia, and an organized group sent the following petition to the state: "Whereas we find that Congress has seaded us to the State of Georgia therefore we think it good to petition the General Assembly of this State to Do to and for us as in their Wisdom think Best." In response to

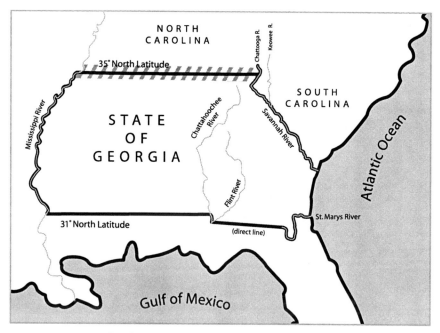

The disputed "nonexistent land," 1787–1811

this plea, Georgia passed a law in 1804 to call the area Walton County, named after George Walton, a signer of the Declaration of Independence from Georgia. Elections were held, and for the next several years Walton County sent representatives to serve in the legislature at the new Georgia capital in Milledgeville. However, at the same time, many residents refused to recognize Walton County as part of Georgia and wanted to be part of North Carolina.

The 35th parallel north was at the heart of the disputed tract of land. North Carolina claimed that Walton County was north of the latitude line; Georgia claimed it was south. In 1808 an abortive attempt to locate 35° north latitude was made by surveyors from both states, but blaming inexperience and poor equipment, they could not agree on its exact location.

In 1810, after years of altercations between citizens loyal to each state, North Carolina dispatched its State Militia to settle the matter. A brief battle occurred near the present town of Brevard, N.C., with several fatalities and prisoners being taken by the militia. Eventually calm was restored, but this event, possibly a combination of fact and

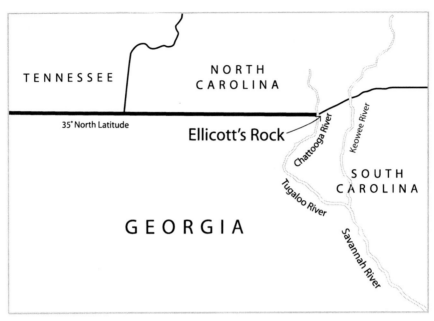

Placement of Ellicott's Rock at 35th parallel, 1811

fantasy, has since been known as the Walton War.

The matter was finally resolved in 1811 when Andrew Ellicott, who performed the United States/Spain border survey, was commissioned by the Governor David B. Mitchell of Georgia to survey and mark the location.

After celebrating the July 4th holiday, Ellicott and one of his sons left Philadelphia and traveled to Savannah, where they met the supplied surveying team. Lacking enough horses, he and the team walked the 200 miles into the woods of north Georgia and North Carolina while the few horses carried their equipment. In late October, after following the Savannah and Chattooga rivers, the team reached the base camp of the 1808 surveying trip in the Appalachian Mountains. Ellicott made his measurements and realized that he was "about 7 miles South and 12 miles West … ." The weather delayed his necessary astronomical observations until Christmas Day in 1811, when he was able to determine the location of the 35th parallel. After chiseling the letters NC and GA into a rock for use as a monument, he placed it on the east side of the Chattooga River.

The monument which Ellicott used to mark the 35th parallel also has a confusing story connected to it. In 1813, two years after Ellicott placed his monument on 35° north, commissioners representing both Carolinas met to document their boundary and confirm Ellicott's measurements. They marked a different rock with "Lat 35 AD 1813 NC+SC" at what they believed to be the junction of the two states, a few feet away from Ellicott's rock. This 1813 marker, rather than the original 1811 monument, is usually called Ellicott's Rock, although it has also been called Commissioner's Rock. The National Register of Historic Places says:

> ... the rock was located and carved in 1813 by the boundary commissioners for the States of North and South Carolina. It was to mark the latitude 35 at the Chattooga River; however, later surveys have located it in 1820 at 34° 59' 02.3", in the 1930s at 35° 00' 04.4" by USGS and by the Tennessee Valley Authority as 35° 00' 02.3". This is the inscription on the rock now known as Ellicott's Rock. The inscription is slowly weathering away due to high rainfall and water action of the Chattooga River.

Since Ellicott's assignment was only to locate and mark the 35th parallel north he considered his job done. For the first time, the northern boundary of Georgia adjacent to North and South Carolina was surveyed and marked.

After walking the 200 miles back to Milledgeville, Ellicott addressed the General Assembly and gave them the news that he had surveyed and found 35° north latitude and that Walton County was not in Georgia but was actually located 18 miles into North Carolina. The governor and legislators were unhappy with the news but accepted it because of Ellicott's reputation. The Assembly, however, ignored Ellicott's repeated request for final payment for his work, although he did receive the huge sum of $4,000 for his expenses.

In May 1812, frustrated with the Georgia General Assembly, Ellicott left Milledgeville to return to Pennsylvania. He worked on his journal of observations for the survey and continued asking for his payment. In August 1813 he wrote, "I have delayed the publication of the journal of our proceedings, with all the astronomical observations, and other scientific operations made use of in determining the boundary between the states of Georgia and N. Carolina ... that I may upon the settlement of the account be enabled to speak as favorable of the government of Georgia as the inhabitants generally." Governor Mitchell's reply was, "Permit me to observe that if the publication of your Journal and the truths it is to contain depend upon the amount of your account, the Government of Georgia disclaims all interest of Concern in it and is perfectly indifferent as to its fate." It appears that Ellicott was never paid for his work and one can conclude that the reason this particular journal, of all the voluminous records of his which survived, cannot be found is that he probably destroyed it.

Alabama, Tennessee, and Montgomery's Corner

The Alabama Territory was divided from Mississippi in 1817 and designated by the U.S. Congress as land being south of Tennessee and west of Georgia. The Tennessee and Georgia legislatures quickly passed resolutions to mark the location of the junction, and both states formed and dispatched survey teams. The Georgia team consisted of commissioner Thomas Stocks, mathematician James Camak, and surveyor Hugh Montgomery. The Tennessee team consisted of commissioner John Cocke and mathematician James Gaines. Tennessee's Governor Joseph McMinn accompanied the teams part of the way but was detained to placate Cherokee chiefs who were angry about the survey.

Since the 1802 Articles of Agreement and Cession defined Georgia's western boundary as being a line from the great bend of the Chattahoochee River and "thence in a direct line to Nickajack, on the Tennessee River ... running up the said Tennessee River and along the Western Bank thereof to the Southern Boundary line of the State of Tennessee," the two teams were to meet at Nickajack, find 35° north latitude, and "plainly mark and designate the same."

Nickajack, an ancient Chickamauga Creek town, is now under Nickajack Lake which was formed by the Nickajack Dam (35° 0' 15" N) originally built in 1913 and reinforced in 1967. Nickajack Cave, located nearby, still exists and is a habitat for the gray bat. During the Civil War, guano from this cave was used as a source for potassium nitrate, or saltpeter, for the Confederation Powder Works in Augusta.

Rather than following the description — crossing the Tennessee River at Nickajack and going to its western bank, then up the river to Tennessee — the teams made some measurements and determined that the 35th parallel north was "two miles south of Nickajack old town," far south of the Tennessee River. Stock's report *Memorandum made during my Tour in running the Dividing Line between Georgia & Tennessee, Commencing 5, May 1818* describes the survey.

> Mr. Gaines took his first observation at 12 o'clock with the Marten equatorial Thiodolite and made this place Lt [latitude] 35. 11. On the 19th, at 12 o'clock Mr. Gaines made his second observation and found this place 11 miles South of the 35th degree differing from his yesterdays observation 22 miles in consequence of which we concluded to dispense with the further use of his instrument. Mr. Camak took his first observation on Spika genus, Lat. 35.2. and continued his observations of Spika and Autares [sic] for two additional days and determined Nickajack to be about 2 miles North of the 35th degree. The star observations continued for several more days until on the 25th Mr. Camak took observation on Antaries [sic], the mean of the above observations makes Camp Cocke south of the 35. Degree. On the 27th the astronomical observatory was moved nearly one-half mile North of Camp Cocke to Camp Montgomery. The next day, Mr. Camak took observations on Spica Zubbenelgenubia, Antaries and Zubenisch [sic]. On the 29th Mr. Camak made observations on each of the above and B. Scorpia. The mean of all the observations at Camp Montgomery make this place Lat. 34.59.13.
> Left Camp Montgomery and measured 72 chains to the 35th degree and planted a Rock marked on the North side Tens. June 1st, 1818 Var 6 3/4 degrees East. On the South side Geo. Lat. 35. No. James Camak. This rock or corner is placed one mile and 7 chains from Tennessee River and about one quarter mile South of Nickajack Cave.

After placing the stone, for some unknown reason, the measurement teams continued beyond their original assignment eastward for 110 miles, where they made a mark at what they thought was the eastern boundary between Georgia and Tennessee. In reality, they were well beyond the Georgia/Tennessee boundary and almost halfway through the Georgia/North Carolina boundary.

The surveying finished, both teams reported to their respective legislatures which then officially recorded their findings. Tennessee's Code of 1819 describes the Georgia/Tennessee boundary as beginning

> ... at a point in the true parallel of the thirty-fifth degree of north latitude, as found by James Camak, mathematician on the part of the State of Georgia and James Gaines, mathematician on the part of the State of Tennessee, on a rock which rock stands one mile and twenty-eight poles from the south bank of the Tennessee River, due south from near the center of the old Indian Town of Nickajack, and near the top of Nickajack mountain, at the supposed corner of the States of Georgia and Alabama. The true dividing line between the states of Tennessee and Georgia is hereby in every part and parcel thereof established as the true southern boundary line of the state of Tennessee. This Act shall take effect and be in force so soon as the state of Georgia shall have passed a law similar in its provisions.

The wording of "the thirty-fifth degree of north latitude, *as found by James Camak*" (emphasis added) haunts Georgia to this day.

Georgia's legislature passed a resolution in 1819 authorizing the governor to have the maps of the surveyed lines recorded in the Surveyor General's Office, but there is no record of any law or act certifying or approving the survey as the official boundary between the two states.

That same year, Georgia asked North Carolina to appoint a team to survey and mark their mutual boundary. James Camak, the mathematician on the Georgia/Tennessee boundary team represented Georgia, along with two surveyors and a commissioner; surveyor Robert Love, Esq. represented North Carolina. Starting at the monument left by Ellicott on the eastern side of the Chattooga River, the Georgia/North Carolina team surveyed 35 miles to the west and came upon the location where the year before the Georgia/Tennessee survey team, coming from Nickajack, had stopped. Unfortunately, the two lines did not connect. The westward line was almost one-half mile north of the termination of the eastward line.

Astonishingly, instead of trying to correct the error, the team simply marked a straight vertical line southward, connecting the two surveyed lines. The offset is called Montgomery's Corner, in honor of Hugh Montgomery, the Georgia surveyor on the 1818 survey team.

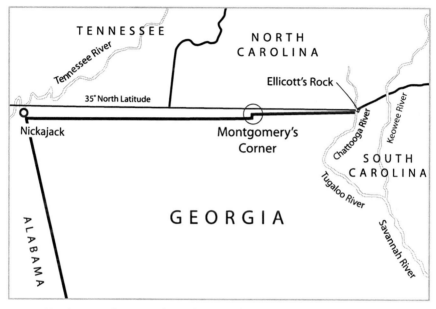

Northern and western boundaries with Montgomery's Corner, 1826

Seven years after Alabama was admitted as the 22nd state, the governors of Alabama and Georgia agreed in 1826 to survey that portion of the boundary line between the two states, which started at Miller's Bend (today's West Point, Ga.) and then northwest to Nickajack as had been provided for in the 1802 land cession.

Georgia appointed a team which consisted of two commissioners, a surveyor, and again James Camak. Alabama appointed two commissioners but did not send a surveyor, believing that it was Georgia's responsibility to mark the line. As the Georgia/Alabama survey team approached Nickajack, Camak began making his measurements for the location of 35° north latitude and immediately noted that this measurement differed from his 1818 measurement.

In his report to Georgia, his location of Nickajack was "about one quarter of a mile north of the Georgia Tennessee boundary as surveyed in 1818." He then made new observations and reported "Latitude of Northern Boundary to be 34° 59' 36" North. Latitude of Nickajack 35° 00' 06" North," which meant that Georgia's northern boundary was not on the 35th parallel north where it was supposed to be but instead was a mile south of it. Addressing the location of the 35° north latitude, which he was supposed to have measured in 1818, Camak wrote, "Admitting that the observations and results of 1826 are correct and that this boundary (Georgia Tennessee) has been fixed 37 and 90/100 chains too far South, the loss sustained by Georgia will equal to 51 51/100 square miles or 33,048 8/10 acres."

Camak offered this excuse for his different measurements:

> In 1826, I used three sextants, one belonging to the state, the other two belonging to the Board of Public Works. I have not much confidence in their accuracy as none of them were made by artists of celebrity. They may answer well enough for marine purposes, but I would never use them on land, if I could will it. The tables I used were good, with the exception of typographical errors; and so far as the calculations are concerned, I can rely, with perfect confidence, on the results. Taking everything into consideration, I am inclined to give a preference to the results of 1826.

James Camak

The difference between the surveys done by Andrew Ellicott and James Camak is profound. Modern techniques have shown all of Ellicott's surveys to be almost perfect, and by Camak's own admission, his measurements are erroneous. Not only was his 1818 survey of the Georgia/Tennessee boundary significantly south of 35° north latitude, his continuing survey of the line going eastward 110 miles was also south of the 35th parallel. His survey in 1819, going westward from Ellicott's Rock measuring the Georgia/North Carolina boundary, is also markedly off-course and well south of the 35th parallel, resulting in Montgomery's Corner. Little did Camak know that had he measured accurately, Chattanooga, Tenn., would be Chattanooga, Ga.

Camak was an interesting Georgian. Although little is known of his early life, he came to Athens in 1817 to teach mathematics at the University of Georgia, and it was there that he met and married the daughter of the president of the school. He only stayed at the university for two years, and it was during this time that he represented Georgia on the survey teams that measured the Georgia/North Carolina and Georgia/Alabama/Tennessee boundaries. Camak eventually moved to Milledgeville and began a textile mill in the 1830s.

Accumulating some wealth, he moved back to Athens where he was the director of the Branch Bank of the State of Georgia and a trustee for the University of Georgia. He was the first president of the Georgia Railroad, the first railroad in the state, and built a large home in Athens which stayed in the family for 100 years. The Camak House is marked with a Georgia historical marker as a landmark in Georgia railroading.

Camak was the first editor of *Southern Cultivator*, a Georgia farm journal, and was involved in starting the county's first agricultural society in 1845. The town of Camak, Ga., was incorporated in 1898 about 40 miles west of Augusta. The date of his death is uncertain.

Florida

The exact location of Georgia's southern boundary gained importance as the state sought to draw settlers to the area after the cession of its western lands. Despite the daunting obstacles presented by the Okefenokee Swamp and the local Creek people, Georgia sent a team of engineers in 1818 to locate Ellicott's Mound and confirm it as the head of the St. Marys River, which formed the eastern part of Georgia's boundary with Spain.

Spanish power in the New World had long been in decline, so when General Andrew Jackson, a national hero as a result of his victory at the Battle of New Orleans in the War of 1812, was given orders in 1817 to prevent Spanish Florida from becoming a refuge for runaway slaves, he overextended his orders and attacked Spanish forts in East Florida. These attacks exposed Spain's military weakness, and Secretary of State John Quincy Adams seized the opportunity to accuse Spain of violating the 1795 Pinckney Treaty, which required that Spain not incite the Native Americans to warfare. Rather than risk a conflict with the United States over their territory in what is now New Mexico, Texas and California, Spain signed the Adams-Onis Treaty in 1819 formally giving all of its Florida territory to the United States. In return, the United States gave up all of its claims to the Spanish-controlled southwestern lands. The United States then sent surveyors to the area to map out land lots for investors.

With Jackson as the military governor, the combined colonies of Florida became a new U.S. territory in 1821, and the surveyor general of the territory sent a surveyor up the St. Marys River to accurately locate Ellicott's Mound. It wasn't until 1845 when Florida became the 27th state that both Georgia and Florida again became interested in accurately measuring their common boundary.

The Trail of Tears

By 1826 all of Georgia's external boundaries with contiguous states had been measured, but within Georgia, the Cherokee nation had its own government and possessed and lived on its own land. Georgia, however, wanted that land available for the thousands of settlers coming into Georgia.

Disregarding the law, Georgia abolished the Cherokee government and established a lottery for distributing its lands to white citizens. In addition, the administration of President Andrew Jackson and a handful of Cherokee signed the Treaty of Echota in 1835 which required the entire Cherokee nation to move west to a reservation in Oklahoma. The Cherokee people refused to go, saying that the treaty was improperly signed, and in 1838 federal troops were brought to Georgia to move them out. Of the 18,000 Cherokee forced from their homes, 4,000 died during the journey, immortalized as the Trail of Tears.

Georgia's boundaries were finally defined, but this was just the beginning of almost two centuries of disagreements over the measured lines. A state's boundary could profoundly affect the citizens whose property was near or straddled the boundary line, and courts would be asked to interpret treaties or boundaries previously agreed to by the states.

Important Dates IV

1795   Yazoo Land Fraud

Pinckney Treaty between United States and Spain

1796   Tennessee becomes 16th state

Andrew Ellicott commissioned to survey United States/Spain border

1798   Third Georgia Constitution adopted

Mississippi Territory designated by United States

1802   Georgia cedes western land to United States

1803   Louisiana Purchase

1810   Walton War

1811   Andrew Ellicott marks 35th parallel north at the Georgia/South Carolina/North Carolina junction

1817   Mississippi becomes 20th state; Alabama Territory designated by United States

1818   Survey of Georgia/Tennessee boundary

First survey of Georgia/Florida boundary

1819   Second survey of Georgia/Florida boundary

First survey of Georgia/North Carolina boundary

Adams–Onis Treaty between United States and Spain

Alabama becomes 22nd state

1821   Florida Territory designated by United States

1825   Third survey of Georgia/Florida boundary

1826   Thomas Jefferson and John Adams die on July 4

First survey of Georgia/Alabama boundary

Second survey of Georgia/North Carolina boundary

Surveying of Georgia's external boundaries complete

1828 Gold discovered in Georgia
1835 Treaty of New Echota between United States and Cherokee
1836 Georgia begins building Western and Atlantic Railroad
1838 Cherokee nation removed from Georgia in "Trail of Tears"
1845 Florida becomes 27th state

Georgia's Boundary Disputes

When the first English settlers came to the New World, they measured their property by metes and bounds as had been done for centuries, using physical features of the local geography, along with directions and distances, to define and describe the boundaries of a parcel of land. However, as landmarks such as trees, streams and rocks change, boundaries begin to blur. Disputes occur because of the passage of time as well as poor wording in documents or faulty measurements. Although early records from all the colonies reflect that property lines were surveyed and measured, every colony on the North American continent questioned the boundaries with their neighbors, and disagreements over the right of navigation on waterways were common.

The use of rivers as boundaries also invite dispute for many reasons other than navigation including accretion resulting in new land mass, erosion of the banks, fishing rights, erecting structures such as docks and marinas, and recreational uses such as swimming and boating. Owners of property next to a river or lake also have certain riparian rights and responsibilities.

Following the Revolutionary War, each of the original 13 colonies acquired title from the Confederation Congress to all the land within their boundaries, and each state's legislature described its boundaries in their statutes. Surveying the boundaries of the new states was very important to the founding fathers, and by the late 1700s, more sophisticated methods of surveying were available. Nonetheless, many surveyors were poorly trained or unable to accurately perform their measurements, so the disputes persisted.

This chapter examines the common boundaries of Georgia and its neighboring states and the lawsuits between them.

The legal boundaries of Georgia

Every state publishes an annual text with the statutes, codes, regulations and laws currently passed or held over from previous years, including a description of their boundaries. (State boundaries, once established, do not change much, so legislators rarely review their boundaries unless an amendment is necessary.) Georgia's statutes are known as the Official Code of Georgia Annotated (O.C.G.A.). This annotated version includes a supplement to each statute which may consist of the legislative history of the statute, summaries of relevant state and federal court cases, the state attorney general's opinions, and references to law review articles. Georgia's code is made up of 44 volumes which contain 53 major subject areas, further subdivided into chapters and sections. The boundary descriptions can be found in Title 50, State Government; Chapter 2, Boundaries and Jurisdiction of the State; Article 1, State Boundaries; Sections 1–5.

O.C.G.A. §50-2-1. Boundaries of the state generally
(2009, amended eleven times since 1851)

> The boundaries of Georgia, as deduced from the Constitution of Georgia, the Convention of Beaufort, the Articles of Cession and Agreement with the United States of America entered into on April 24, 1802, the Resolution of the General Assembly dated December 8, 1826, and the adjudications and compromises affecting Alabama, Florida, and South Carolina are as follows:
>
> From the sea, at the point where the northern edge of the navigable channel of the River Savannah intersects a point three geographical miles east of the ordinary low water mark, generally along the northern edge of the navigable channel up the River Savannah, along the northern edge of the sediment basin to the Tidegate, thence along the stream thereof to the fork or confluence made by the Rivers Keowee and Tugalo, and thence along said River Tugalo until the fork or confluence made by said Tugalo and the River Chattooga, and up and along the same to the point where it touches the northern boundary line of South Carolina, and the southern boundary line of North Carolina, which is at a point on the thirty-fifth parallel of north latitude, reserving all the islands in said Rivers Savannah, Tu-

galo, and Chattooga, except for the Barnwell Islands and Oyster Bed Island in the Savannah, to Georgia; thence on said line west, to a point where it merges into and becomes the northern boundary line of Alabama — it being the point fixed by the survey of the State of Georgia, and known as Nickajack; thence in a direct line to the great bend of the Chattahoochee River, called Miller's Bend — it being the line run and marked by said survey; and thence along and down the western bank of said Chattahoochee River, along the line or limit of high-water mark, to its junction with the Flint River; thence along a

The boundaries of Georgia, 2009

certain line of survey made by Gustavus J. Orr, a surveyor on the part of Georgia, and W. Whitner [sic], a surveyor on the part of Florida, beginning at a fore-and-aft tree about four chains below the junction; thence along this line east, to a point designated 37 links north of Ellicott's Mound on the St. Marys River; thence along the middle of said river to the Atlantic Ocean, and extending therein three geographical miles from ordinary low water along those portions of the coast and coastal islands in direct contact with the open sea or three geographical miles from the line marking the seaward limit of inland waters; thence running in a northerly direction and following the direction of the Atlantic Coast to the point where the northern edge of the navigable channel of the River Savannah intersects a point three geographical miles east of the ordinary low water mark, the place of beginning; including all the lands, waters, islands, and jurisdictional rights within said limits; and also all the islands within three geographical miles of the seacoast.

This description of Georgia's general boundaries is *de jure*, meaning "according to the law," or what they should be in principle, and for over 200 years, Georgia's legislators have continued to define the boundaries in this manner. However, the citizens of Georgia would be better served if the boundaries were defined as *de facto*, meaning "in actual practice," because the boundaries have changed over the years due to faulty surveying, environmental changes, lawsuits or simply because the legislators remain unaware.

The current (and all of Georgia's previous) legal boundary descriptions are erroneous for the following reasons:

1. The northern boundary of Georgia is described as being where the Chattooga River crosses the 35th parallel and then "on said line west" until it meets the northern boundary of Alabama. In fact, the northern boundary of Georgia does not continue west on 35° north latitude because of the faulty surveys of 1818, 1819 and 1826. All published maps of Georgia clearly show the northern boundary as being below 35° north.

2. The description of the Georgia/Alabama boundary — "the point fixed by the survey of the State of Georgia, and known as

Nickajack" — is inaccurate since there is no such place as Nickajack, a Native American village in the 18th century, now located somewhere under Nickajack Lake, Tenn.
3. The western boundary of Georgia is "from Nickajack thence in a direct line to the great bend of the Chattahoochee River, called Miller's Bend." Miller's Bend is no longer in existence but is likely West Point, Ga.
4. The western boundary is further described as "down the western bank of said Chattahoochee River, along the line or limit of high-water mark, to its junction with the Flint River." The actual boundary goes down the western bank of the Chattahoochee to 31° latitude north, then continues down the *middle* of the river — not the high-water mark — to its junction with the Flint River as stipulated in the 1763 Treaty of Paris and further defined by the Georgia Supreme Court in 1930 (discussed on p. 99).
5. The junction of the Flint and Chattahoochee Rivers is unknown, being somewhere beneath Lake Seminole which was constructed in the 1950s.
6. Georgia's southern boundary "beginning at a fore-and-aft tree about four chains below the junction" of the Chattahoochee and Flint rivers is archaic and impossible to determine since this location is also now under Lake Seminole.

It is obvious that Georgia legislators have not looked carefully at this portion of the O.C.G.A. in quite some time. While trying to find the exact location of the ancient village of Nickajack which is now under a lake may not be practical, describing the boundary line on the western side between Georgia and Florida and between Georgia and Alabama is a simple correction requiring a mere two or three sentences. The *de jure* and *de facto* boundary descriptions should be the same; legislators need to revisit the topic and make the appropriate changes to properly and officially describe the boundaries of Georgia as they exist in the real world.

Resolving disputes between states

As part of organizing the states into one cohesive federal union and anticipating that states themselves would become litigants, the drafters of the U.S. Constitution of 1787 were farsighted enough to require lawsuits between states to be heard in the U.S. Supreme Court rather than at a local level. The Constitution provides in Article I, Section 10, that, "No State shall, without the Consent of Congress, enter into any Agreement or Consent with another State." In addition, Article III, Section 2, states, "In all Cases in which a State shall be Party, the Supreme Court shall have original Jurisdiction."

Original jurisdiction means that, rather than its customary role as an appellate court hearing cases that have already been heard by a lower court and jury, the Supreme Court will hear suits between two states first, as the judge and jury, then render an opinion. Often the Supreme Court appoints a special master, usually a prominent jurist with experience or special knowledge of the subject, to regulate the proceedings, hear the evidence and make a conclusion.

Georgia and Florida boundary disputes

O.C.G.A. §50-2-5. Boundary between Georgia and Florida
(2009, amended eight times since 1859)

> The boundary line between Georgia and Florida shall be the line described from the junction of the Flint and Chattahoochee Rivers to the point 37 links north of Ellicott's Mound, on the St. Marys River; thence down said river to the Atlantic Ocean; thence along the middle of the presently existing St. Marys entrance navigational channel to the point of intersection with a hypothetical line connecting the seawardmost points of the jetties now protecting such channel; thence along said line to a control point of latitude 30 degrees 42' 45.6" north, longitude 81 degrees 24' 15.9" west; thence due east to the seaward limit of Georgia as now or hereafter fixed by the Congress of the United States; such boundary to be extended on the same true 90 degrees bearing so far as a need for further delimitation may arise.

Florida Statutes §6.09: Boundary between Florida and Georgia

> (1) The line run and marked by B.J. [sic] Whitner, Jr., on the part of Florida, and G.J. Orr, on the part of Georgia, is the permanent

boundary line between the States of Florida and Georgia.

(2) The boundary line between the States of Florida and Georgia as described in subsection (1) herein shall be extended from a point 37 links north of Ellicott's Mound on the Saint Marys River; thence down said river to the Atlantic Ocean; thence along the middle of the presently existing Saint Marys entrance navigational channel to the point of intersection with a hypothetical line connecting the seaward-most points of the jetties now protecting such channel; thence along said line to a control point of latitude 30° 42' 45.6" N., longitude 81° 24' 15.9" W.; thence due east to the seaward limit of Florida as now or hereafter fixed by the Congress of the United States; such boundary to be extended on the same true 90° bearing so far as a need for further delimitation may arise.

Astonishingly, both the O.C.G.A. and the Florida statute completely leave out a description of the Georgia/Florida boundary along the Chattahoochee River from 31° north latitude to its junction with the Flint River.

This portion of the boundary has rarely been in dispute, except for a Georgia Supreme Court case in 1930. Florida Gravel Co. v. Capital City Sand & Gravel Co. (170 Ga. 855, 1930) concerned the removal of sand, gravel and minerals from the east bank of the Chattahoochee River in Seminole County, Ga., in the southwestern corner of the state at the junction of the Chattahoochee and Flint rivers. Because the law is applied differently depending on the type of river involved, the court first discussed the three types of rivers — private, navigable and tidal — and determined that the Chattahoochee River was navigable.

The court's opinion began by citing the 1802 Agreement and Cession of Lands between Georgia and the United States of America, which provided that Georgia's western boundary was the *western* bank of the Chattahoochee River beginning at 31° north latitude and going up to Nickajack at 35° north. However, according to the Pinckney Treaty of 1795 between the United States and Spain, the western boundary of Georgia began at 31° north and then went down the *middle* of the Chattahoochee River to its junction with the Flint River. The court opined that Georgia's western boundary line was different depending on whether the line was north or south of the 31st parallel.

The court held that the Georgia/Florida boundary south of 31° north was in the middle of the Chattahoochee River, and since the gravel was being removed from the east bank south of the 31st parallel, the Capital City Sand & Gravel Co. had title to the land to the middle of the river and therefore had the right to remove the sand, gravel and minerals from the river bed.

Since this 1930 decision, there have been no cases between Georgia and Florida disputing their boundary in the middle of the river. No significant commercial traffic occurs on the Chattahoochee, although Lake Seminole, at the junction of the Flint River, provides valuable recreational opportunities.

The history of the Georgia/Florida boundary going eastward from this junction to the Atlantic Ocean, however, is a fascinating and complex story. In 1799 when Andrew Ellicott surveyed the United States/Spain border, the southern boundary of Georgia was also the southern border of the United States. Subsequent to Ellicott's measurement of the border, numerous attempts were made to ascertain the true line, with each measurement being different from the others.

Shortly after being admitted as a state, Florida filed a lawsuit against Georgia regarding their joint boundary (Florida v. Georgia, 58 U.S. 478, 1854). Both states had surveyed the eastern part of the boundary in the previous decades, and the United States had also surveyed this area after Spain had given East and West Florida to the United States in the Adams-Onis Treaty. All three entities had sold parcels of land along what each considered the boundary line.

Georgia claimed that the boundary began at the junction of the Chattahoochee and Flint rivers and continued in a straight line eastward to Lake Spalding (also called Lake Randolph), located some 30 miles south of Ellicott's Mound. Florida, using plats surveyed by the United States as well as plats published by its own surveyor, D.F. McNeil, maintained that Ellicott's Mound was accurately located at the head of the St. Marys River along what was referred to as the

McNeil Line.

The issue of who was correct about the location of Ellicott's Mound was never addressed by the Supreme Court because it first had to consider motions brought by Georgia, Florida and the United States. Georgia moved that the court appoint a surveyor to make the appropriate measurements; Florida asked the court to "take out commissions to examine witnesses in the case and for sundry orders to expedite the case and prepare it for trial." The United States asked the court for permission to be a party to the lawsuit and to be allowed to present their evidence as to the location of the boundary.

After a lengthy and eloquent discussion, a majority of the divided court delivered the opinion denying all three of these motions. The court ordered both Florida and Georgia to prepare their own cases with their own documents and witnesses and to be ready for trial in one year, i.e., December 1855. However, no further hearings were ever conducted in the U.S. Supreme Court as Florida and Georgia soon agreed to reappraise their positions and accepted Ellicott's Mound as the true and accurate location of the mouth of the St. Marys River.

An 1887 U.S. Supreme Court case, Coffee v. Groover (123 U.S. 1, 1887), concerned a boundary dispute between Florida and Georgia 40 years earlier which resulted in title to property being given to two different parties. While the resultant decision of who owned the property is not germane to the states' boundary dispute, the convoluted chronology of the dispute is artfully discussed in the court's narrative and deserves being repeated here.

> In early colonial times, there were always mutual complaints of encroachment between the British provinces and the Spanish province of Florida, sometimes resulting in military conflicts, and no boundary was ever settled between them. The difficulty was finally removed by the treaty of 1763, by which Florida was ceded to Great Britain. ... Soon after this event, on the seventh of October, 1763, King George III, by proclamation, erected governments in the newly acquired territories of Canada and the Floridas, and established the boundaries of

the latter as follows, to-wit: "The government of East Florida, bounded to the westward by the Gulf of Mexico and the Appalachicola River, *to the northward by a line drawn from that part of said river where the Chattahoochee and Flint Rivers meet to the source of the St. Mary's River*, and by the course of the said river to the Atlantic Ocean." West Florida was bounded north by the parallel of 31 north latitude, from the Mississippi to the Chattahoochee River. ... The above-defined line, from the junction of the Chattahoochee and Flint Rivers to the source of the St. Mary's, has from 1763 to the present time been the recognized boundary line between Georgia and Florida. The land in controversy is situated about midway between its extremities.

By the definitive treaty of peace with Great Britain in 1783, the line above described was adopted as the southern boundary line of the United States, and the Floridas were at the same time ceded to Spain. ... By the Treaty of October 27, 1795, between the United States and Spain, this boundary was confirmed, and it was provided that a commissioner and a surveyor should be appointed by each party to meet at Natchez within six months from the ratification of the treaty and proceed to run and mark the boundary line, and make plats, and keep journals of their proceedings, which should be considered as part of the treaty. Our government appointed Andrew Ellicott, Esq., as commissioner in May, 1796, and a surveyor to assist him, and they proceeded to Natchez, and after much procrastination on the part of the Spanish authorities, a Capt. Stephen Minor was appointed on the part of Spain; and the joint commissioners of the two countries, in 1798 and 1799, ran and marked the boundary line from the Mississippi to the Chattahoochee, and determined the geographical position of the junction of the Chattahoochee and Flint Rivers to be in north latitude 30° 42' 42.8" and west longitude 85° 53' 15". The hostility of the Creek Indians prevented them from running the line east of the Chattahoochee; but they sailed around the coast of Florida and up the river of St. Mary's, and fixed upon the eastern terminus of the straight line prescribed in the treaties at the head of the St. Mary's, where it issues from the Okefenoke Swamp, and erected a mound of earth to designate the spot. This was in February, 1800.

The mound is still in existence, and is called "Ellicott's Mound," and appears on all the principal maps of that part of the country. The commissioners, supposing that the true head of the river was located in the swamp, agreed that it should be considered as distant two miles northeast from the mound, and that in running the boundary line from the Chattahoochee, it should be run to the north of the mound, and not nearer to it than one mile. The point fixed upon as the head of the St. Mary's was determined by observations to be in north latitude 30° 21' 39 1/2" west longitude 82° 15' 42". The distance by straight line or great circle from the junction of the Chattahoochee and Flint Rivers to the head of the St. Mary's was calculated at 155.2 miles; and the initial course for running the line from each terminus was given, with the proper corrections to be made at intervals in order to follow the great circle. The commissioners signed a joint report of their proceedings, and transmitted the same to their respective governments. All these particulars are set forth in Mr. Ellicott's journal, and are matters of public history. ...

It thus appears that by authority of the United States and Spain, the *termini* of the line in question were fixed and settled in February, 1800. It only remained for any competent surveyor to follow the directions of the commissioners in order to trace the actual boundary line on the ground. The country in the region traversed by this line was occupied, in the early part of the century, by the nation of Creek Indians, and there was no immediate demand for having it run and marked. And, as under the Constitution, no state could enter into a treaty with the Indians, it became the interest of Georgia to make some arrangement with the government of the United States to take measures for the gradual removal of Indian occupancy. A convention was accordingly entered into between Georgia and the United States on the twenty-fourth of April, 1802, by which the former ceded to the latter all her territory between the Chattahoochee and the Mississippi Rivers, and the United States ceded to Georgia all their right to any public lands south of Tennessee and the Carolinas and east of the Chattahoochee not within the proper boundaries of any state, and agreed to extinguish the Indian title within the State of Georgia as early as could be peaceably done. ...

The state being now desirous of disposing of her lands and introducing settlers thereon, naturally turned her attention to the question of the true location of the boundary line between her own territory and that of the Spanish province of Florida. Some person, professing to be better than others as to the topography of the country about the head of St. Mary's River, asserted that the commissioners, Ellicott and Minor, in seeking its source, had ascended the wrong branch — namely the north branch — whereas the true St. Mary's, or main stream came from the west and took its source many miles further south than the point fixed upon by them. The Legislature of Georgia took up the matter, and in December 1818, the Senate passed a resolution requesting the Governor to appoint proper persons to proceed without delay to ascertain the true head of St. Mary's River, and if it should appear that the mound thrown up by Ellicott and Minor was not at the place set forth in the treaty with Spain, that they make a special report of the facts and that the Governor communicate the same to the President of the United States with a request that the lines might be run agreeably to the true intent and meaning of the treaty. ... In pursuance of this request, the governor appointed three eminent engineers, Gens. Floyd, Thompson, and Blackspear, to make the examination suggested, and immediately, by a letter dated February 17, 1819, communicated the fact to the executive government at Washington. The engineers made a careful *reconnaissance* of the country about the head streams of the St. Mary's, accompanied by the person who had made the supposed discovery, and became satisfied that his information was at fault, and reported that after a careful examination they found the head of the river to agree with the report made by Mr. Ellicott. This result was also communicated to the Executive at Washington, and thus ended for the time being the claim on the part of Georgia to have the eastern terminus of the boundary line readjusted and changed. Soon after this proceeding, in 1819, the state employed one J.C. Watson to run and mark the line. This is the origin of the line called "Watson's Line;" and to this line the state laid out its counties and townships, surveyed its public lands, and made grants to settlers. But it nowhere appears that this line ran to Ellicott's Mound or near to it; on the contrary, it would seem from other conceded facts that it ran considerably south of it. As we have already seen, the lands in controversy in the present case adjoin this line, being situated on the north side of it.

Florida was ceded to the United States in 1819, and possession of the territory was taken by Gen. Jackson in July, 1821. In 1825, the surveyor general of the government for the Territory of Florida, preparatory to a survey of the public lands therein, caused the boundary line between Georgia and Florida to be run out and marked by D. F. McNeil, a deputy surveyor, and the line so run was called "McNeil's Line." At the point in controversy, which (as before said) is about midway between the two extremities of the straight line called for by the treaty, it ran, according to the testimony, 14 chains to the north of Watson's Line; but how near it approached Ellicott's Mound at the eastern extremity does not appear. The government surveys in Florida were made to bound on this line, and of course overlapped, more or less, the Georgia surveys and grants extending to Watson's Line.

The State of Georgia, about this period, perhaps in consequence of the location of McNeil's Line, by a communication of her Governor to the government of the United States, requested that joint measures should be undertaken for a mutual and final settlement of the boundary. The matter being referred to Congress, an act was passed on the fourth of May, 1826, by which the President was authorized, in conjunction with the constituted authorities of the State of Georgia, to cause to be run and distinctly marked the line dividing the Territory of Florida from the State of Georgia from the junction of the Rivers Chattahoochee and Flint to the head of St. Mary's River, and for that purpose to appoint a commissioner or surveyor, or both: "provided that the line so to be run and marked shall be run straight from the junction of said Rivers Chattahoochee and Flint to the point designated as the head of St. Mary's River by the commissioners appointed under the third article of the treaty [with Spain] made October 27, 1795." ... This act, it will be seen, adopted the eastern terminus of the line as settled by Ellicott and Minor. The President thereupon appointed the Ex-Gov. Thomas M. Randolph, of Virginia, as commissioner under the act, and the executive of Georgia appointed Thomas Spaulding; and the commissioners entered upon their joint duties in February, 1827, and appointed John McBride as their common surveyor. They continued their operations for over two months; but the Georgia commissioner having, as he supposed, notwithstanding the report of the commissioners of 1819, discovered that the western branch of the St. Mary's River was the largest and longest stream, and therefore the true river, the governor of the state suddenly brought the survey to a close by recalling the assent of Georgia and withdrawing the powers of her commissioner. ... From this time onward for many years, a controversy was carried on between Georgia, on the one side, and the United States and Florida, on the other with regard to this boundary line; Georgia contending that the line should be run to Lake Randolph, the head of the western or southern branch of the St. Mary's, and the United States and Florida contending that it should run to the head of the northern branch, as settled and determined by the commissioners, Ellicott and Minor, under the treaty. ...

In 1845 Florida was admitted into the union as a state, embracing all the territories of East and West Florida, as ceded by Spain to the United States by the Treaty of 1819 [sic]. ... Renewed efforts were soon afterwards made by Florida and Georgia to effect a settlement of the boundary, but without success. In 1850 the State of Florida filed a bill in this court against the State of Georgia to procure a determination of the controversy. In December term, 1854, the Attorney General was allowed to intervene on the part of the United States. ... Evidence was taken by the parties, but in consequence of the war and the final settlement of the controversy by mutual agreement, the cause was never brought to a hearing. In 1857 the governors of the two states had a conference which resulted in an agreement by which Georgia relinquished her pretensions to have the eastern terminus of the line changed; and the *termini* fixed by the commissioners, Ellicott and Minor, were substantially adopted. The following resolutions and enactments of the legislatures of the two states will show the course of negotiation, and the terms of the arrangement finally concluded between them.

On the twenty-fourth of December, 1857, the following resolution was adopted by the legislature of Georgia, to-wit:

"Whereas in the matter of controversy now pending in the supreme court of the United States between the state of Florida and the state of Georgia touching the boundary line of the two states, we deem it of much importance that this protracted and expensive litigation should cease; and whereas, with a view to the settlement of the question, a negotiation has been progressing between the late executives of the aforesaid states the result of which was an agreement to adopt the terminal points of the present recognized line as the true terminal points of the boundary line, to be resurveyed, corrected, and marked, provided it is shown by either party that the present line is incorrect, the agreement aforesaid being made subject to the ratification of the legislatures of the two states: Resolved, (1) that we do hereby ratify the action of the late executive of this state in accepting the proposition of the governor of Florida to adopt the terminal points of the present recognized line as the true terminal points of the boundary line, and will regard, adopt, and act upon the present line, as run and recognized between those points, as the settled boundary of the two states, or will so recognize and adopt any other line between those points, which may be ascertained and established on a resurvey and remaking of the boundary: provided said boundary correction is made by virtue of law and by joint action of the states aforesaid. (2) Be it further resolved by the authority aforesaid that should it be deemed essential or important by either state to have the boundary line between the terminal points of the present recognized boundary resurveyed and remarked, the governor of this state is hereby authorized to appoint a competent surveyor to join any such surveyor appointed on the part of Florida to run out and mark distinctly such a line from one to the other terminal point herein indicated, to be known as the line and settled boundary between the two states; the surveyor on the part of Georgia to be paid such compensation as may be determined on by the present or any future legislature. (3) And be it further resolved that the governor of this state shall, so soon as the same shall have passed both branches of the present general assembly, transmit a certified copy to the governor of Florida.

"Approved December 24, 1857."

This resolution was responded to by the legislature of Florida on the twelfth of January, 1859, by passing a resolution in precisely the same terms, *mutatis mutandis*; and on the fifteenth of the same month, an act was passed by the legislature of Florida for bringing into market, as soon as the line should be settled, all state lands bordering thereon that had not been disposed of, giving to the occupants whose right was not disputed five months to purchase the lands occupied by them at their appraised valuation. As one or both of the parties desired to have a resurvey made between the terminal points, the state of Georgia appointed George F. [sic] Orr, and the state of Florida B.F. Whitner, surveyors to run and mark the line accordingly. They commenced their work in 1859, and it is referred to in the subsequent acts and resolutions.

An act was passed by the legislature of Georgia on the sixteenth of December, 1859, referring to the fact that the joint surveyors were running their first trial line, and agreeing to adopt it as conclusive if Florida would do the same: provided that on the eastern terminus it did not depart exceeding one-fourth of a mile from Ellicott's Mound, but that if it was not accepted by Florida, and if therefore a new line would have to be run so as to get a straight line from the mouth of Flint River to Ellicott's Mound, then the line thus designated and marked by the surveyors should be the permanent boundary between the two states. The act also proposed the passage of laws to quiet the titles of *bona fide* holders of lands under grants of either Georgia or the United States. The response made by the legislature of Florida to this proposition was the passage of an act on the twenty-second of December, 1859, substantially adopting the proposition made by Georgia, declaring "that the line now being run by B.F. Whitner, Jr., on the part of Florida, and G. J. Orr, on the part of Georgia, be, and the same is hereby recognized and declared to be the permanent boundary line between the two states so soon as the same shall be permanently marked by said surveyors, provided that said line at its eastern terminus does not depart from or miss Ellicott's Mound more than one-fourth of a mile, or twenty chains;" and declaring secondly "that the titles of *bona fide* holders of land under any grant from the State of Georgia which land may fall within this state by the foregoing line are hereby confirmed and conveyed to said holders so far as any right may accrue to this state: provided nothing herein shall apply to lands to which citizens of this state may claim title south of what is known as the 'McNeil Line.'" It turned out that the line run by Orr and Whitner ran even further north than the McNeil Line; but it came within the stipulated distance from Ellicott's Mound, namely within a quarter of a mile; in fact, within 37 links, or less than 25 feet, north of the mound. ... This was more favorable to Georgia than the line agreed on by Ellicott and Minor, which was to run at least one mile north of the mound.

On the 14th of December, 1860, the legislature of Georgia, probably considering that its last proposition was not fully accepted, passed a resolution directing the governor to reopen negotiations with the authorities of Florida in regard to the boundary line and to urge its so as to protect the rights of citizenship and the titles of lands held under grants from Georgia and if practicable so as to retain and keep the fractional lots sold by Georgia within the jurisdiction of the state. In response to this resolution, the Florida Legislature, on the eighth of February, 1861, passed the following resolution, to-wit: "Whereas, [by] an act approved by the governor, twenty-second December, 1859, it was by the general assembly enacted that the line then being run by B.F. Whitner, Jr., on the part of Florida, [and] G.J. Orr, on the part of Georgia, should be and was thereby recognized and declared to be the permanent boundary line between the states of Georgia and Florida as soon as the same should be permanently marked by said surveyors: provided the said line at its eastern terminus did not depart from or miss Ellicott's Mound more than one-fourth of a mile or twenty chains; and whereas, the said line has been run and marked by said surveyors on the part of the two states, the eastern terminus of which, so run and marked, is within the distance prescribed in said proviso: therefore resolved that the line run and marked by B.F. Whitner, Jr., on the part of Florida, and G.J. Orr, on the part of Georgia be and the same is hereby declared to be the permanent boundary line between the two States of Georgia and Florida, and that the governor be, and he is hereby, requested to issue his proclamation that the said line so run and marked has been and is declared to be the permanent boundary line between the two states: provided the state of Georgia shall have on its part declared the said line to be the boundary between that state and Florida. Be it further resolved that the governor be requested to forward a copy of these resolutions to the governor of Georgia with a request that similar steps be taken by Georgia, so that the question of boundary may be finally settled." …

By a long and argumentative resolution passed by the legislature of Georgia on the eleventh of December, 1861, after stating the respective positions taken by the two states, it was proposed as follows:

"The general assembly, to avoid further dispute, proposes to her sister state Florida that what is denominated the 'Watson Line' (which will leave in the limits of this state the fractional lots of land heretofore sold under an act of her legislature) shall be adopted as the boundary line. The settlement upon this basis will not interfere with the rights of citizenship as claimed by the citizens of either state." Florida made no answer to this proposition.

Finally, by a resolution passed on the thirteenth of December,

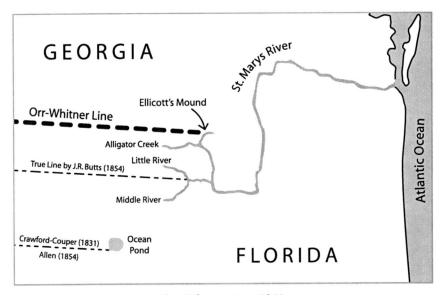

Orr–Whitner Line, 1861

1866, the legislature of Georgia, referring to the act of sixteenth December, 1859, and recognizing the fact that the Orr and Whitner Line as run did not depart exceeding one-fourth of a mile from Ellicott's Mound, and referring also to the action of the Florida Legislature of February 8, 1861, adopted the Orr and Whitner Line as "the permanent boundary line between the States of Georgia and Florida." And this agreement thus finally arrived at by the two states was recognized and confirmed by an act of Congress approved April 9, 1872, entitled "An act to settle and quiet the title to lands along the line between the states of Georgia and Florida," by which it declared "that the titles to all lands lying south of the line dividing the States of Georgia and Florida known as the 'Orr and Whitner Line,' lately established as the true boundary between said states, and north of the line run by Georgia, known as the 'Watson Line,' being all the lands lying between said lines, be, and the same are hereby, confirmed, so far as the United States has title thereto, in the present owners deriving titles from the state of Georgia."

The Orr–Whitner line, surveyed in 1859, was accepted by both Florida and Georgia in 1861 as the official boundary between the two states, although there is no physical evidence of the measurement of the line. The long debate of Georgia's southern boundary with Florida was finally settled and has not been disputed since.

Georgia and Alabama boundary disputes

O.C.G.A. §50-2-4. Boundary Between Georgia and Alabama
(2009, amended six times since 1863)
> The boundary line between Georgia and Alabama shall be the line described from Nickajack to Miller's Bend on the Chattahoochee River, and down said river to its junction with the Flint River.

Code of Alabama, §41-2-2. Boundary between Alabama and Georgia
(1975, amended eight times since 1852)
> The boundary line between Alabama and Georgia commences on the west side of the Chattahoochee River, at the point where it enters

the State of Florida; from thence up the river, along the western bank thereof, to the point on Miller's Bend, next above the place where the Uchee creek empties into such river; thence in a direct line to the Nickajack.

As already discussed, Nickajack and Miller's Bend are no longer in existence and the use of these historical names is inaccurate. While Alabama's description is, for the most part, correct (although specifying 31° north latitude instead of "where it enters the State of Florida" would be more precise), Georgia's description incorrectly extends the boundary to the junction of the Chattahoochee and Flint rivers, approximately 20 miles south. Below 31° north, Georgia's boundary is with Florida. Additionally, the language "down said river" is vague and should state that the Georgia/Alabama boundary line is "along the western bank of the Chattahoochee River."

The first challenge to the boundary line came in 1845, not from either state but from two of their citizens, Stephen M. Ingersoll of Alabama and John H. Howard of Georgia. There was no procedural rule concerning lawsuits between citizens of two different states as there is today, so Ingersoll and Howard each filed a suit in their respective state. Eventually, through the appellate system, the cases arrived at the U.S. Supreme Court which combined the two into a single case. The Supreme Court's opinion in Howard v. Ingersoll (54 U.S. 381, 1851) affected the Georgia/Alabama boundary more than either of the original cases.

Ingersoll filed a lawsuit against Howard in the Circuit Court of Alabama claiming that when Howard build a dam on the Chattahoochee River to provide water for the mill on his property in Georgia, the water backed up and overflowed onto Ingersoll's land in Alabama, causing damage to his mill. The Alabama jury returned a verdict in favor of Ingersoll for the sum of $4,000. Howard appealed to the Superior Court of Alabama, where the judgment was affirmed, and in 1850, he appealed to the U.S. Supreme Court.

Almost simultaneously in Georgia, Howard, as the plaintiff, filed his case against Ingersoll claiming that when Ingersoll's land was flooded in high water, Ingersoll was trespassing by fishing in the water on property belonging to Howard, since the boundary to Howard's property in Georgia extended to the west bank of the Chattahoochee and would include the water in the river. The Georgia jury returned a verdict for Howard for the sum of $600.

The Supreme Court's opinion runs for dozens of pages and reviews the history of Georgia and Alabama starting with colonial Georgia in 1733. The most important points are these:

> This case involves a question of much higher interest and importance than a simple decision upon the rights of these parties, as the Court sees that the decision cannot be reached without a determination of the boundary-line between two sovereign States for a distance of some one hundred and fifty miles. The facts in the record are few, being confined to a description of the localities respecting this boundary at the point in dispute, and the few that are disclosed very imperfectly and confusedly stated. It is to be regretted that the court is obliged to pass upon a question of this magnitude under these embarrassments and in the absence of any opportunity on the part of the two States interested to furnish the necessary topographical information in respect to the River Chattahoochee and its western banks for the whole distance within which they constitute the boundary between them. ...
>
> Grants of land, bounded by the sea or by navigable rivers, where the tide ebbs and flows, extend to high-water mark, that is, to the margin of the periodical flow of the tide, unaffected by extraordinary causes, and the shores below common high-water mark belong to the State in which they are situated. But grants of land bounded on rivers above tide-water, or where the tide does not ebb and flow, carry the grantee to the middle of the river, unless there are expressions in the terms of the grant, or something in the terms taken in connection with the situation and condition of the lands granted, that clearly indicate an intention to stop at the edge or margin of the river. There must be a reservation or restriction, express or necessarily implied, which controls the operation of the general presumption, and makes the particular grant an exception.
>
> ... And this raises the material and important question in the case, namely, where shall that line be drawn? On behalf of Georgia, it

is contended, it shall be drawn on the bank or bluff, as described in the record, at high-water mark; on behalf of Alabama, at the bank or ridge of sand and gravel, where the western margin of the river is found at ordinary low-water mark.

In our judgment, the true boundary-line intended by Georgia and the United States, and the one fairly deducible from the language of the cession [1802 Articles of Agreement and Cession], is the line marked by the permanent bed of the river by the flow of the water at its usual and accustomed stage, and where the water will be found at all times in the season except when diminished by drought or swollen by freshets. This line will be found marked along its borders by the almost constant presence and abrasion of the waters against the bank. It is always manifest to the eye of any observer upon a river, and is marked in a way not to be mistaken. The junction of bank and water at this stage of the river satisfies the words of the cession, and furnishes a line as fixed and certain as is practicable; and is just and reasonable to all the parties concerned.

The Court also discussed the riparian rights of those whose property is next to a body of water with the following:

Every proprietor of land on the banks of a river has naturally an equal right to the use of the water which flows in the stream adjacent to his lands. No proprietor has a right to use the water to the prejudice of other proprietors, above or below, unless he has acquired a prior right to divert it. He has no property in the water itself, but a simple usufruct while it passes along. Any one may reasonably use it who has a right of access to it; but no one can set up a claim to an exclusive right to the flow of all the water in its natural state; and that what he may not wish to use himself shall flow on till lost in the ocean. Streams of water are intended for the use and comfort of man; and it would be unreasonable, and contrary to the universal sense of mankind, to debar a riparian proprietor from the application of the water to domestic, agricultural, and manufacturing purposes, provided the use works no substantial injury to others.

The Supreme Court reversed the findings from each state and sent the cases back to those lower courts with directions to follow its opinion regarding each case. Since Howard v. Ingersoll was a private issue between two citizens, not between two states, the legal description of the Georgia/Alabama boundary was not changed. The issue did not surface again until 1855 when Alabama filed a lawsuit against Georgia in the U.S. Supreme Court.

Alabama did not send a surveyor on the joint Georgia/Alabama team that in 1826 had marked the line from Miller's Bend on the Chattahoochee River to Nickajack, but then the state protested to Georgia for over a decade about the accuracy of the survey. In its suit, Alabama asked the court to define the boundary on the Chattahoochee River where the low-water mark met the western bank. Georgia insisted that the 1802 Land Cession Agreement described the western boundary of Georgia as the western bank, meaning the location where the high-water mark met the western bank, and therefore Georgia had jurisdiction over all the soil and land up to the high-water mark on the west bank of the river.

In 1859 the U.S. Supreme Court rendered its opinion by reviewing the history of the treaties and agreements which had defined the boundary, as well as providing various definitions for the words "bank," "channel," "high water," "low water," "shore" and "river." Alabama v. Georgia (64 U.S. 505, 1859) specifically rejected Alabama's low-water mark as the boundary line and concluded that

> ... by the contract of cession, Georgia ceded to the United States all of her lands west of a line beginning on the western bank of the Chattahoochee river where the same crosses the boundary line between the United States and Spain, running up the said Chattahoochee river and along the western bank thereof.
>
> We also agree and decide that this language implies that there is ownership of soil and jurisdiction in Georgia in the bed of the river Chattahoochee, and that the bed of the river is that portion of its soil which is alternately covered and left bare, as there may be an increase or diminution in the supply of water, and which is adequate to contain it at its average and mean stage during the entire year, without reference to the extraordinary freshets of the winter or spring, or the extreme droughts of the summer or autumn.
>
> The western line of the cession on the Chattahoochee river must be traced on the water-line of the acclivity [upward slope] of the

western bank, and along that bank where that is defined; and in such places on the river where the western bank is not defined, it must be continued up the river on the line of its bed, as that is made by the average and mean stage of the water, as that is expressed in the conclusion of the preceding paragraph of this opinion.

By the contract of cession, the navigation of the river is free to both parties.

This decision seemed to satisfy all the parties, and there have been no other lawsuits between the two states about their boundary. The importance of what is low water or high water, or how far up the bank the boundary goes, has diminished over the years as the commercial and navigable use of the river has markedly declined.

Georgia and Tennessee boundary disputes

O.C.G.A. §50-2-3. Boundary between Georgia and North Carolina and Tennessee (2009, amended six times since 1863)

The boundary between Georgia and North Carolina and Georgia and Tennessee shall be the line described as the thirty-fifth parallel of north latitude, from the point of its intersection by the River Chattooga, west to the place called Nickajack.

Tennessee Code Annotated §4-2-105. Georgia boundary. (2008)

The boundary line between this state and the state of Georgia begins at a point in the true parallel of the thirty-fifth degree of north latitude, as found by James Carmack [sic], mathematician on the part of the state of Georgia, and James S. Gaines, mathematician on the part of this state, on a rock about two feet (2') high, four inches (4") thick, and fifteen inches (15") broad, engraved on the north side thus: "June 1st, 1818, Var. 6¾ East," and on the south side thus: "Geo. 35 North, J. Carmack," which rock stands one (1) mile and twenty-eight (28) poles from the south bank of the Tennessee River, due south from near the center of the old Indian town of Nick-a-Jack, and near the top of the Nick-a-Jack Mountain at the supposed corner of the states of Georgia and Alabama; thence running due east, leaving old D. Ross two (2) miles and eighteen (18) yards in this state, and leaving the house of John Ross about two hundred (200) yards in the state of Georgia, and the house of David McNair one (1) mile and one fourth (¼) of a mile in this state, with blazed and mile-marked trees, lessening the variation of the compass by degrees, closing it at the termination of the line on the top of the Unicoi Mountain at five and one half degrees (5½°).

Of all the issues involving Georgia's boundaries, none is more significant than the northern boundary with Tennessee. This boundary should be located on the 35th parallel of north latitude but was erroneously surveyed several times and is actually about one mile below 35° north. Virtually insurmountable legal hurdles would need to be cleared in order to change the boundary line since Georgia has never asked the U.S. Supreme Court for a decision.

After the 1818 measurement of the Georgia/Tennessee line, Tennessee's legislature passed a resolution approving the boundary and added, "Be it enacted, that this act shall take effect and be in force so soon as the state of Georgia shall have passed a law similar in its provisions." Georgia responded by authorizing the Governor to "have recorded in the surveyor general's office of this state, the maps of the lines as run, dividing this state and the states of Tennessee and North Carolina, with the certificates thereunto annexed and pay for the same out of the contingent fund." Camak's map of the Georgia/Tennessee line is recorded in the Surveyor General's Office, now part of the Georgia Archives, but there is nothing in either the House Journal or Senate Journal acknowledging that the legislature or the governor ever recognized this boundary as the legal boundary of the state.

Georgia's land lotteries and railroads played a large part in the history of the northern boundary with Tennessee and North Carolina. Shortly after the 1802 cession of lands to the United States, Georgia began a series of land lotteries to distribute land within the new boundary to its white citizens. (Georgia was the only state to use lotteries for land distribution, and today almost 75% of all private land in Georgia is a result of the early 1800s lotteries.) Because the underlying reason for this method of land distribution was to remove the Creek and Cherokee people from Georgia, the state did not intend to profit from the lotteries; fees were very low and depended on the land lot size. White males of legal age, widows, and orphans could all apply. Lots along Georgia's northern boundary were surveyed before giving title to the new owner, yet the deeds were based on the erroneous surveys by Camak. Georgia made no complaints about the survey, and all the northern counties of Georgia bordering on Tennessee were said to be "within the limits of Georgia."

In 1836 Georgia began building the state-owned Western and Atlantic Railroad — the same railroad made famous by the Great Locomotive Chase during the Civil War — from Atlanta to Chattanooga along the Tennessee River. To do so, Georgia needed permission to cross the border and the General Assembly passed an act in November 1836 to send a representative to Tennessee for "authorizing the extension of our State Road from the Georgia line through the Territory of Tennessee to the Tennessee River." Tennessee granted the extension in January 1838, and construction of the railroad from the Georgia line to the Tennessee River continued. Again, Georgia did not raise a concern regarding the boundary.

Fifty years later, in October 1887, Georgia's legislature passed an act stating there was some doubt as to the location of the state line between Georgia and Tennessee which runs between Dade County in Georgia and Marion and Hamilton counties in Tennessee and authorizing the governor to communicate with the governor of Tennessee to commission a joint survey of this part of the line. The Tennessee legislature responded by passing an act in April 1889 to provide for the establishment of line between Georgia and Tennessee.

> Whereas, There are grave doubts as to the location of the state line between Georgia and Tennessee, on that part of the line between Dade county, Georgia, and Marion and Hamilton counties, Tennessee, creating trouble and inconvenience between the citizens of the two states; therefore,
>
> Section I. Be it enacted by the general assembly of the State of Tennessee, That the governor of this state shall be empowered to communicate with the governor of Georgia for the purpose of having a joint survey looking to the settlement of the question in dispute.

> Sec. 2. That the governor of Tennessee, after such communication with the governor of Georgia as is necessary, shall be empowered to appoint three competent men to act with such members as may be appointed by the governor of Georgia, whose duty it shall be to survey, establish, and proclaim the true line between the disputed points.
>
> Sec. 3. That not exceeding two hundred and fifty dollars he, and the same is hereby, appropriated to pay expenses of said proceeding, for which the governor of Tennessee may draw his warrant on the comptroller.
>
> Sec. 4. That the joint commission so appointed by the governors of Tennessee and Georgia shall begin their survey at that point where Georgia and Alabama corner, and run east as far as is necessary to establish the disputed line.

Unfortunately both Georgia and Tennessee failed to follow up or act in any way to resurvey the line. Thus began more than a century of unsuccessful attempts to resolve the boundary issue.

In July 1891 William Northen, the new governor of Georgia, wrote to Tennessee Governor John P. Buchanan enclosing a copy of Georgia's 1887 Act and asking him to respond. Hearing nothing for six months, Northern again wrote Buchanan, who again ignored him. Peter Turney took office as Tennessee's governor in January 1893, and in June, Northern wrote to the new governor, asking for his cooperation in resurveying the boundary. Turney, following the tradition of his predecessor, also ignored Northern. In September Northern again wrote Turney asking him for a response and again received none.

Four years later, in December 1897, the Georgia legislature passed a resolution stating

> ... there was great uncertainty concerning the true northern boundary of this State, And, whereas, the States of North Carolina and Tennessee have each recognized that grave doubt exists as to the true boundary lines between those States respectively and the State of Georgia and have each by Acts of their respective General Assemblies provided for the appointment of commissioners to confer with such commissioners as may be appointed by the State of Georgia for

the purpose of ascertaining the true boundary line.

W.A. Wimbish, an attorney for the Western and Atlantic Railroad who had prepared a report to the state regarding property rights of the railroad in Chattanooga because of Tennessee's increasingly hostile conduct, was chosen to render a report to the legislature regarding the boundary. The Wimbish Report of 1898 was a recitation of Georgia and Tennessee's resolutions regarding the boundary, as well as the attempts by Georgia to communicate with Tennessee.

Armed with this information, the Georgia legislature passed a law in December 1898 creating a tribunal to be composed of nine commissioners — three each from Georgia, North Carolina and Tennessee — to act as a court of original jurisdiction to hear evidence and make a decision regarding the accuracy of the boundary. The Georgia legislature included a caveat that if neither Tennessee or North Carolina adopted similar legislation, then Georgia's governor would take no action. Again Tennessee, and now North Carolina, did not respond.

Georgia: Comprising Sketches of Counties, Towns, Events, Institutions and Persons in Cyclopedic Form quotes *Appleton's Annual Cyclopedia for 1898* as stating:

> Georgia and Tennessee have a boundary dispute which involves the position of the city of Chattanooga. The present boundary places the city in Tennessee but several expert geographers have recently found information which goes to show that the boundary line is not properly located. The boundary line between the states is the thirty-fifth parallel of north latitude, and this was located by a survey in 1818 at a point one mile south of the Tennessee River. Georgia will claim that a correct survey will place the thirty-fifth parallel to be north of Lookout Mountain and that more than 100,000 citizens of Tennessee will have to become citizens of Georgia.

The 1905 address of Tennessee Governor James Frazier to his legislators recalled the issue of the boundary and the 1889 Tennessee law which had been passed regarding the appointment of a commission to survey the line involving Dade, Marion and Hamilton counties. The Tennessee legislature responded by passing a law that year almost identical to the previous law, including the requirement that the governor appoint a commission to resurvey the line, except for the requirement that the survey begin at the Georgia/Alabama/Tennessee corner. Even though the legislature did revisit the issue as he had requested, Frazier failed to follow through during the duration of his governorship and no commission was appointed.

In August 1906, ignoring the Tennessee resolution of 1905, Georgia passed another resolution.

> Since there is a dispute as to the location of the line between the State of Georgia and the State of Tennessee, and especially as to that part of the line between the counties of Fannin in Georgia and Polk in Tennessee [at the eastern end of the boundary, not at the western end where Dade, Marion and Hamilton counties are located] exist, the governor should therefore confer with the Governor of Tennessee and take such steps as are necessary to settle and locate the line between the said States and counties.

Nothing was said about appointing a commission, and nothing became of the resolution.

Governor Tom C. Rye of Tennessee wrote Governor John M. Slaton of Georgia in March 1915 suggesting that the boundary line needed to be fixed. Slaton replied that he would be in Chattanooga to attend a meeting where Rye would also be present, and they could discuss the matter. There is no record of any meeting between the two.

In August 1916 Georgia passed yet another resolution

> ... calling for a resurvey of the eastern end of the boundary between the County of Fannin in the State of Georgia and County of Bek [sic] in the State of Tennessee. The Governor is authorized to take the necessary steps with the proper authorities looking to the establishment of the true boundaries between the counties herein named and the proper marking of the same.

Georgia now seemed more focused on the eastern part of the boundary. Regardless, Tennessee as usual ignored the resolution and Georgia as usual did nothing.

Another six years passed, and in 1922 the Georgia legislature passed this resolution:

> Whereas there is a dispute as to the location of the line between the State of Georgia and the State of Tennessee, and especially to that

> part of the line between the counties of Walker and Dade in Georgia, and Hamilton and Marion in Tennessee, therefore, the Governor is requested to confer with the Governor of Tennessee and take such steps as are necessary to settle and locate the line.

And for the 10th time since 1818, both states failed to address the accuracy of their common boundary.

Georgia's legislature addressed the boundary line issue once more in March 1941 and this time appointed

> ... a joint standing committee of three Senators and five Representatives to meet with a committee which the Tennessee legislature would appoint, to establish, survey and proclaim the true boundary line between Georgia and Tennessee and to take such further or other action of pursue such remedy or remedies as a majority of the committee might deem proper.

Georgia Governor Eugene Talmadge sent the signed legislation to Tennessee's governor but received no reply.

The Georgia legislature passed two resolutions in 1947 related to the boundary; the first provided for the appointment of a committee to negotiate with a Tennessee committee, and the second authorized the attorney general to institute proceedings in the federal courts if no settlement could be obtained by direct negotiations. Tennessee Governor Jim McCord did hold a conference with the Georgia delegation, but no decision was reached and the attorney general did not pursue a lawsuit.

Another 25 years passed before Georgia reconsidered the boundary issue. In 1971 the Georgia General Assembly enacted a resolution authorizing Governor Jimmy Carter to communicate with the governors of Tennessee and North Carolina to settle the dispute. The resolution also called for the creation of the Georgia/North Carolina and Georgia/Tennessee boundary line commission to meet with similar commissions from North Carolina and Tennessee to take whatever actions were necessary to resolve the boundary matter. There is no record that Carter wrote to Tennessee's governor, or had any response from him, or that any commissions ever met with either state.

The drought of 2008 served as a catalyst to re-examine the boundary and brought this issue to the headlines of newspapers and television. If the 35th parallel of north latitude had been properly surveyed, the Georgia/Alabama/Tennessee boundary at the northwest corner of the state would be in the middle of the Tennessee River, and Georgia would have plenty of water for its citizens. For the 14th time since 1818, the Georgia General Assembly addressed the issue, passing a joint House and Senate resolution, signed by Governor Sonny Perdue in May 2008, which recounts the history of the dispute and concludes with these words:

> Now, therefore, be it resolved by the General Assembly of Georgia that the Governor of Georgia has the full support of the General Assembly and is hereby strongly urged to initiate negotiations with the Governors of Tennessee and North Carolina for the purpose of correcting the flawed 1818 survey erroneously marking the 35th parallel south of its actual location and to officially recognize the State of Georgia's northern border with the States of Tennessee and North Carolina as the precise 35th parallel as was intended when both states were created. The Governor of Georgia shall have the authority to negotiate settlement of the issue for the State of Georgia which shall be binding upon the State with the approval of such agreement by the General Assembly.
>
> Be it further resolved that should the Governor of Georgia's negotiations with the Governors of Tennessee and North Carolina fail to come to any resolution to the issue of the disputed boundary between the two states, then the Attorney General of Georgia is authorized to take the appropriate legal action to correct Georgia's northern border at the 35th parallel. Such legal action by the Attorney General of Georgia includes, but is not limited to, initiating suit in the United States Supreme Court against either or both of the States of Tennessee and North Carolina for final settlement of this boundary issue.
>
> Be it further resolved that it is the clear and express intent of the General Assembly to correct, establish, survey, and proclaim the northern border of the State of Georgia and the southern border of the States of Tennessee and North Carolina at the true 35th parallel.

The Tennessee General Assembly responded on April 10, even before Georgia's governor had signed the Georgia Joint Resolution, with Tennessee House Joint Resolution 919, "A resolution relative to the Tennessee-Georgia boundary."

> Whereas, action recently taken by the Georgia General Assembly calls into question the border between the State of Tennessee and the State of Georgia; and
>
> Whereas, the Georgia General Assembly has passed legislation claiming a boundary dispute between Georgia and Tennessee at the 35th Parallel and proposing to settle such dispute by the creation of a Boundary Line Commission composed of legislators from both States; and
>
> Whereas, the Georgia General Assembly claims that erroneous surveys conducted in 1818 and 1826 have deprived Georgia of access to the Tennessee River; and
>
> Whereas, this General Assembly realizes that the Tennessee-Georgia boundary has been well established for nearly 200 years, and that there is no valid reason for Tennessee to revisit this issue; and
>
> Whereas, in addition to the doctrine of adverse possession, in which long-term possession of real property trumps survey boundaries, all other pertinent legal precedent favors the State of Tennessee; and
>
> Whereas, the United States Supreme Court, the highest court in the land, has held in Oklahoma vs. Texas that there is a "general principle of public law" that, as between States, a "long acquiescence in the possession of territory under a claim of right and in the exercise of dominion and sovereignty over it, is conclusive of the rightful authority" and has held in Georgia vs. South Carolina that "long acquiescence in the practical location of an interstate boundary, and possession in accordance therewith, often has been used as an aid in resolving boundary disputes" between States; and
>
> Whereas, this General Assembly understands that original jurisdiction in boundary disputes between the several States of this great nation resides with the United States Supreme Court; and

Whereas, this General Assembly is presently considering substantive measures to address Tennessee's water supply and water shortages; and

Whereas, it is this General Assembly's duty to protect the borders and waters of our State for present and future generations of Tennesseans; now, therefore

Be it resolved by the House of Representatives of the one hundred fifth general assembly of the State of Tennessee, the Senate concurring, that on behalf of the State of Tennessee and all Tennesseans, this General Assembly respectfully declines to participate in the Boundary Line Commission proposed by the Georgia General Assembly, or any similar commission established for such purpose. Be it further resolved, that it is the sense of this General Assembly that the Tennessee-Georgia boundary has been legally established since 1818.

Be it further resolved, that enrolled copies of this resolution be delivered to the Speaker of the House of Representatives and the President of the Senate of the Georgia General Assembly.

Even though it is quite clear that it was surveyed and marked erroneously, Georgia's northern boundary with Tennessee and North Carolina will probably never be changed. Georgia's decades of huffing and puffing instead of asking for the court's assistance has left the state with three seemingly insurmountable legal obstacles to overcome: laches, prescription and acquiesence. Laches can be defined as a failure to do what the law requires, a procrastination or "sleeping on one's rights," and can be shown when the defendant — in this case, Tennessee — would be in a worse position now if the plaintiff prevailed than it would have been at the time the claim should have been brought. The 35th parallel north, which is where the boundary was intended to be, runs through the city of Chattanooga, Tenn., and relocating the boundary there now would mean that thousands of people would suddenly be citizens of Georgia, clearly to the detriment of Tennessee. The doctrine of prescription is a legal way of acquiring property by long-term, open, continuous and peaceful use, which both Tennessee and North Carolina have demonstrated. Finally, acquiesence means that Georgia's silence and failure to file a lawsuit against Tennessee or North Carolina implies Georgia's consent for them to use and own the land.

As of the publication of this book, to the best of the author's knowledge after consulting with the Office of the Attorney General of the State of Georgia, no action has been taken by Georgia regarding the filing of a lawsuit with the U.S. Supreme Court against the states of Tennessee and North Carolina.

Georgia and North Carolina boundary disputes

O.C.G.A. §50-2-3. Boundary between Georgia and North Carolina and Tennessee (2009, amended six times since 1863)
> The boundary between Georgia and North Carolina and Georgia and Tennessee shall be the line described as the thirty-fifth parallel of north latitude, from the point of its intersection by the River Chattooga, west to the place called Nickajack.

Chapter 141 of the statutes of the North Carolina General Assembly, entitled "State Boundaries," consists of eight sections, none of which describe the North Carolina/Georgia boundary. Only one even mentions Georgia.

N.C. Gen. Stat. §141-1. Governor to cause boundaries to be established and protected. (2008)
> The Governor of North Carolina is hereby authorized to appoint two competent commissioners and a surveyor and a sufficient number of chainbearers, on the part of the State of North Carolina, to act with the commissioners or surveyors appointed or to be appointed by any of the contiguous states of Virginia, Tennessee, South Carolina, and Georgia, to return and remark, by some permanent monuments at convenient intervals, not greater than five miles, the boundary lines between this State and any of the said states.
>
> The Governor is also authorized, whenever in his judgment it shall be deemed necessary to protect or establish the boundary lines between this State and any other state, to institute and prosecute in

the name of the State of North Carolina any and all such actions, suits, or proceedings at law or in equity, and to direct the Attorney General or such other person as he may designate to conduct and prosecute such actions, suits, or proceedings.

The lengthy and confusing *Revised Statutes of The State of North Carolina passed by the General Assembly at the Session of 1836–37*, located in the Legislative Library of the North Carolina General Assembly, illustrates the convoluted history of the boundary.

Whereas the States of Georgia and North Carolina, by their respective commissioners duly authorized for that purpose, did, on the eighteenth day of June, in the year of our Lord one thousand eight hundred and seven, (1807) at Buncombe courthouse, enter into articles of conventional agreement, as follow:

Art. I. It is mutually agreed and admitted, the territories of the said States of Georgia and North Carolina, as far as they adjoin each other, are, and of right, ought to be, separated and bounded by the thirty fifth degree of north latitude; and for the purposes of preventing in future all manner of dissensions concerning jurisdiction, the underwritten commissioners will proceed forthwith to ascertain the said thirty fifth degree of north latitude and to run and mark the line accordingly; which line, when ascertained and competed, with joint concurrence, shall forever after be regarded as the line of separation and boundary between the two states.

Art. II. The commissioners on the part of Georgia do not consider their powers competent to enter into any stipulations which would bind the government of the said state to confirm entries or grants for land heretofore made or obtained under the authority of the State of North Carolina, which land, on the running of the line, may be found to be within the State of Georgia; but, impressed with the justice of a certain proportion of the said claims, and the peculiar circumstances which entitle them to consideration, the said commissioners promise and agree to recommend them in a special manner to the liberality of the government, not doubting but that the legislature whereof will, by law, provide for the confirmation and establishment of the said titles, in a manner which will afford a satisfactory and adequate relief. And to this end, the said commissioners will recommend the establishment of an impartial tribunal for the special purpose of inquiring into and ascertaining the various descriptions of such claims and of determining on each according to their respective merits, and as reason and equity may require; which tribunal the said commissioners will also recommend to be composed of three persons to be appointed and paid by each state; but they shall convene and hold their meetings in the state of Georgia, and their decisions shall be conclusive.

Art. III. There having been great dissensions between the people resident in the neighboring counties of Buncombe and Walton, and the said dissensions having produced many riots, routs, affrays, assaults, batteries, trespasses, wounds and imprisonment's, as well on the one side as on the other, and it being of primary importance that peace and tranquility would be restored, and all animosity and ill-will forever buried between the people, who, from their local situations, will, in all probability, be constrained to continue in the vicinity of each other; and as the several outrages committed on both sides proceeded more (as the undersigned are impressed) from themselves constitutionally bound, than from a wish to injure their neighbors or disturb the public peace, the undersigned agree to recommend, in the most ernest manner, to the legislatures of their respective states, to pass laws of amnesty, forgiveness and oblivion for all such offenses, (under the degree of capital) as may have been committed within the said counties of Buncombe and Walton respectively, subsequent to the tenth day of December, in the year one thousand eight hundred and three (1803) and which shall have arisen from and had relation to the disputes which existed concerning the jurisdictions of the two states.

And whereas the said commissioners, with like authority, did on the 27th day of June, in the year aforesaid, at Douthard's Gap, enter into articles in addition and supplementary to the convention agreed on between the commissioners of Georgia and North Carolina, at Buncombe court house, on the eighteenth day of June, in the year aforesaid, which articles are as follow:

The commissioners of the states of Georgia and North Carolina having discovered, by repeated astronomical observations made on the Blue Ridge, and elsewhere, that the thirty fifth degree of north latitude is not to be found on any part of said ridge of mountains east of the line established by the general government as the temporary boundary between the white people and the Indians; and having no authority to proceed over that boundary for the purpose of ascertaining the said thirty fifth degree of north latitude and of running and marking the line accordingly, and being desirous that all causes of collision and irritation between the jurisdictions and people of the two states may be effectually and completely prevented, have agreed to the following articles in addition and supplementary to the convention agreed to at Buncombe court house on the eighteenth day of the present month, viz;

Art. I. The commissioners of Georgia, for and on the part of their state, acknowledge and admit, which acknowledgment and admission are founded on the aforesaid astronomical observations, that the state of Georgia hath no claim to the soil or jurisdiction of any part of the territory north or west of the ridge of mountains which divides the eastern from the western waters, commonly called the Blue Ridge, and east or south of the present temporary boundary line between the white people and the Indians; and that they will consequently recommend to the legislature of the state of Georgia to repeal, at their next ensuing session, the act to establish the county of Walton, and to abrogate and annul all executive and ministerial or other proceedings for the organization thereof.

Art. II. The commissioners on the part of the state of North Carolina, promise and agree to recommend to their government, and particularly to the magistrates, sheriffs and other officers, civil and military, in the county of Buncombe, to execute the laws concerning forfeitures and penalties and in any other respect where the state may be concerned (under the degree of felony) up and towards the people who have adhered to the state of Georgia in the late dissensions concerning jurisdiction, with mildness and clemency; and if the said officers can do it consistently with their obligations of official duty that they forbear to institute suits and to distrain or execute for forfeitures and penalties incurred as aforesaid, between the tenth day of December, in the year eighteen hundred and three (1803) and the date of this agreement until the sense of the legislature shall be had and known thereon.

In order, therefore, that said conventional agreement and the articles additional and supplemental thereto, may be carried into full and complete effect: Be it enacted, That the said conventional agreement and the articles, additional and supplemental thereto, and all and every article and clause thereof, be and the same are hereby fully ratified and confirmed.

An Act to Confirm the Boundary Line between this State and the State of Georgia so far as the Same Has Been Run (passed in the year 1819)

Whereas the states of Georgia and North Carolina, by their respective commissioners, duly authorized for that purpose, have run and marked in part, the boundary line between the said states in conformity with articles of conventional agreement made and concluded by and between the said states, by their respective commissioners, at Buncombe court house, on the eighteenth of June, one thousand eight hundred and seven (1807):

And whereas the said first mentioned commissioners have reported the running and marking said boundary line as follows: To commence at Ellicott's rock and run due west on the thirty fifth degree of north latitude and marked as follows: the trees on each side of the line with three chops, the fore and aft trees with a blaze on the east and west side, the mile trees with the number of miles from Ellicott's rock, on the east side of the tree and a cross on the east and west side; whereupon the line was commenced under the superintendence of the undersigned commissioners jointly: Timothy Tyrral, Esquire, surveyor on the part of the commissioners of the state of Georgia and Robert Love, Esquire, surveyor on the part of the commissioners of the state of North Carolina — upon which latitude the undersigned caused the line to be extended just thirty miles due west, marking and measuring as above described, in a conspicuous manner throughout; in addition thereto, they caused at the end of the first eleven miles after first crossing the Blue Ridge, a rock to be

set up descriptive of the line, engraved thereon upon the north side,
September 25th 1819, N.C. and upon the south side, 35 degree
N.L.G.; then after crossing the river Cowee or Tennessee, at the end
of sixteen miles, near the road, running up and down the said river, a
locust post, marked thus, on the south side, Ga. October 14, 1819;
and on the north side, 35 degree N.L.N.C., and then at the end of
twenty one miles and three quarters, the second crossing of the Blue
Ridge, a rock engraved on the north side, 35 degree N.L.N.C. and on
the south side, Ga. 12th Oct 1819; then on the rock at the end of the
thirty miles, engraved thereon, upon the north side, N.C.N.L. 35
degree G. which stands on the north side of a mountain, the waters
of which fall into Shooting creek, a branch of the Highwasse, due
north of the eastern point of the boundary line, between the states of
Georgia and Tennessee, commonly called Montgomery's Line, just six
hundred and sixty one yards.

1. Be it enacted, That the said boundary line, as described in the
said report, be, and the same is hereby fully established, ratified and
confirmed, forever, as the boundary line between the state of North
Carolina and Georgia.

2. And be it further enacted That this act shall be in force from and
after the passing thereof.

The preceding language in the North Carolina statutes was written in 1836 and has not been changed since that time. There have been no lawsuits between Georgia and North Carolina since the survey of the boundary was made in 1819.

Georgia and South Carolina boundary disputes

O.C.G.A. §50-2-2. Boundary of Georgia and South Carolina
(2009, amended seven times since 1863)

The boundary between Georgia and South Carolina shall be the line
described as running from the mouth of the River Savannah, up said
river and the Rivers Tugalo and Chattooga, to the point where the
last-named river intersects with the thirty-fifth parallel of north latitude, conforming as much as possible to the line agreed on by the

commissioners of said states at Beaufort on April 28, 1787, except for the Barnwell Islands and the Oyster Bed Island in the River Savannah; provided, however, that the boundary along the lower reaches of the Savannah River, and the lateral seaward boundary, shall be more particularly described as being:

BEGINNING at a point 32 degrees 07 minutes 00 seconds North Latitude and 81 degrees 07 minutes 00 seconds West Longitude, located in the Savannah River, and proceeding in a southeasterly direction down the thread of the Savannah River equidistant between the banks of the River on Hutchinson Island and on the mainland of South Carolina, including the small downstream island southeast of the aforesaid point, at ordinary stage, until reaching the vicinity of Pennyworth Island;

Proceeding thence easterly down the thread of the northernmost channel of the Savannah River as it flows north of Pennyworth Island, making the transition to the said northernmost channel using the triequidistant method between Pennyworth Island, the Georgia bank on Hutchinson Island, and the South Carolina mainland bank, thence to the thread of the said northernmost channel equidistant from the South Carolina mainland bank and Pennyworth Island at ordinary stage, around Pennyworth Island;

Proceeding thence southeasterly to the thread of the northern channel of the Savannah River equidistant from the Georgia bank on Hutchinson Island and the South Carolina mainland bank, making the transition utilizing the triequidistant method between Pennyworth Island, the Georgia bank on Hutchinson Island, and the South Carolina mainland bank;

Proceeding thence southeasterly down the thread of the Savannah River equidistant from the Hutchinson Island and South Carolina mainland banks of the river at ordinary stage, through the tide gates, until intersecting the northwestern (farthest upstream) boundary of the "Back River Sediment Basin," as defined in the "Annual Survey — 1992, Savannah Harbor, Georgia, U.S. Coastal Highway, No. 17 to the Sea," U.S. Army Corps of Engineers, Savannah District, as amended by the Examination Survey — 1992 charts for the Savannah Harbor Deepening Project, Drawings No. DSH 112/107, (hereinafter the

"Channel Chart");

Proceeding thence along the said northwestern boundary to its intersection with the northern boundary of the Back River Sediment Basin, in a generally southeasterly direction until said boundary intersects the northern boundary of the main navigational channel as depicted on the Channel Chart at the point designated as SR-34 (Georgia State Grid, East Zone, 1927 NAD, coordinates x=849479.546, y=759601.757);

Proceeding thence toward the mouth of the Savannah River along the northern boundary of the main navigational channel at the new channel limit as depicted on the Channel Chart, via Oglethorpe Range through point SR-33 (coordinates x=853126.849, y=761229.575), Fort Jackson Range through point SR-32 (coordinates x=854568.183, y=762555.255), the Bight Channel through points SR-31 (coordinates x=855854.367, y=765145.946), SR-30 (coordinates x=857363.583, y=766237.604), SR-29 (coordinates x=858471.561, y=766530.527), SR-28 (coordinates x=859881.928, y=766491.887), and SR-27 (coordinates x=861359.826, y=765804.794), Upper Flats Range through point SR-26 (coordinates x=863655.959, y=763821.629), Lower Flats Range through points SR-25 (coordinates x=865361.347, y=759910.744), SR-24 (coordinates x=866413.099, y=758260.171), SR-23 (coordinates x=867339.230, y=757647.194), SR-22 (coordinates x=870024.011, y=756511.390), and SR-21 (coordinates x=873855.646, y=755906.677), Crossing Range through points SR-20 (coordinates x=875581.821, y=754992.833), and SR-19 (coordinates x=884667.253, y=744780.789) and New Channel Range around the Rehandling Basin, and along the northern boundary of the Oyster Bed Island Turning Basin through point SR-16 (coordinates x=894907.977, y=742529.752), to the easternmost end of Oyster Bed Island at Navigational Buoy R "24";

Proceeding thence from Navigational Buoy R "24" easterly along the mean low water line of Oyster Bed Island to the point at which the mean low water line of Oyster Bed Island intersects the Oyster Bed Island Training Wall;

Proceeding thence along the southern edge of the Oyster Bed Island Training Wall until reaching the Jones Island Range line;

Proceeding thence southeasterly along the Jones Island Range line until reaching the northern boundary of the main navigational channel as depicted on the Channel Chart;

Proceeding thence southeasterly along the northern boundary of the main navigational channel as depicted on the Channel Chart to Navigational Buoy R "6," via Jones Island Range and Bloody Point Range; and finally

Proceeding thence in an easterly direction from Navigational Buoy R "6" in a straight line forming the seaward lateral boundary line to the seaward limit of Georgia as now or hereafter fixed by the Congress of the United States, said boundary line bearing approximately 104 degrees from magnetic north, the bearing of said line being more particularly described as being at right angles to the baseline from the southernmost point of Hilton Head Island and the northernmost point of Tybee Island, drawn by the Baseline Committee in 1970.

Provided, however, that the boundary shall be as more particularly shown by reference to the United States Department of Commerce, National Oceanic and Atmospheric Administration (NOAA) GPS coordinates on a map to be prepared by NOAA as a part of the survey commissioned by the States of Georgia and South Carolina in order to locate this boundary. In case of any conflict between the verbal description set forth hereinabove and the map locating the boundary with reference to GPS points, the location shown on the map shall prevail.

Provided, further, that nothing herein shall in any way be deemed to govern or affect in any way the division between the states of the remaining assimilative capacity, that is, the capacity to receive wastewater and other discharges without violating water quality standards, of the portion of the Savannah River described herein.

South Carolina Code of Laws (unannotated), §1-1-10. Jurisdiction and boundaries of the State. (2008)

... From the state of Georgia, this State is divided by the Savannah River, at the point where the northern edge of the navigable channel of the Savannah River intersects the seaward limit of the state's territorial jurisdiction; thence generally along the northern edge of the navigable channel up the Savannah River; thence along the northern edge of the sediment basin to the Tidegate; thence to the confluence of the Tugaloo and Seneca Rivers; thence up the Tugaloo River to the confluence of the Tallulah and the Chattooga Rivers; thence up the Chattooga River to the 35th parallel of north latitude, which is the boundary of North Carolina, the line being midway between the banks of said respective rivers when the water is at ordinary stage, except in the lower reaches of the Savannah River, as hereinafter described. And when the rivers are broken by islands of natural formation which, under the Treaty of Beaufort, are reserved to the state of Georgia, the line is midway between the island banks and the South Carolina banks when the water is at ordinary stage, except in the lower reaches of the Savannah River, as hereinafter described.

The boundary between Georgia and South Carolina along the lower reaches of the Savannah River, and the lateral seaward boundary, is more particularly described as follows and depicted in "Georgia–South Carolina Boundary Project, Lower Savannah River Segment, Portfolio of Maps" prepared by the United States Department of Commerce, National Oceanic and Atmospheric Administration, National Ocean Service, National Geodetic Survey, Remote Sensing Division–2001 (copies available at the South Carolina Department of Archives and History and the South Carolina Geodetic Survey):

Beginning at a point where the thread of the northernmost branch of the Savannah River equidistant between its banks intersects latitude 32° 07' 00" N., (North American Datum 1983–86), located in the Savannah River, and proceeding in a southeasterly direction down the thread of the Savannah River equidistant between the banks of the Savannah River on Hutchinson Island and on the mainland of South Carolina including the small downstream island southeast of the aforesaid point, at ordinary stage, until reaching the vicinity of Pennyworth Island;

Proceeding thence easterly down the thread of the northernmost channel of the Savannah River known as the Back River as it flows north of Pennyworth Island, making the transition to the said northernmost channel using the equidistant method between Pennyworth Island, the Georgia bank on Hutchinson Island, and the South Carolina mainland bank, thence to the thread of the said northernmost channel equidistant from the South Carolina mainland bank and Pennyworth Island at ordinary stage, around Pennyworth Island;

Proceeding thence southeasterly to the thread of the northern channel of the Savannah River equidistant from the Georgia bank on Hutchinson Island and the South Carolina mainland bank, making the transition utilizing the equidistant method between Pennyworth Island, the Georgia bank on Hutchinson Island, and the South Carolina mainland bank;

Proceeding thence southeasterly down the thread of the Savannah River equidistant from the Hutchinson Island and South Carolina mainland banks of the river at ordinary stage, through the tide gates, until reaching the northwestern (farthest upstream) boundary of the "Back River Sediment Basin", as defined in the "Annual Survey–1992, Savannah Harbor, Georgia, U. S. Coastal Highway, No. 17 to the Sea", U. S. Army Corps of Engineers, Savannah District as amended by the Examination Survey–1992 charts for the Savannah Harbor Deepening Project, Drawings No. DSH 1 12/107 , (hereinafter the "Channel Chart");

Proceeding thence along the said northwestern boundary to its intersection with the northern boundary of the Back River Sediment Basin; thence southeasterly until said northern boundary intersects the northern boundary of the main navigational channel as depicted on the Channel Chart at the point designated as SR-34 (latitude 32° 05' 01.440" N., longitude 081° 02' 17.252" W., North American Datum (NAD 1983–86);

Proceeding thence toward the mouth of the Savannah River along the northern boundary of the main navigational channel at the new channel limit as depicted on the Channel Chart, via Oglethorpe Range through point SR-33 (latitude 32° 05' 17.168" N., longitude 081° 01' 34.665" W., NAD 1983–86), Fort Jackson Range through point SR-32 (latitude 32° 05' 30.133" N., longitude 081° 01' 17.750" W., NAD 1983–86), the Bight Channel through points SR-31 (latitude 32° 05' 55.631" N., longitude 081° 01' 02.480" W., NAD 1983–86),

SR-30 (latitude 32° 06' 06.272" N., longitude 081° 00' 44.802" W., NAD 1983–86), SR-29 (latitude 32° 06' 09.053" N., longitude 081° 00' 31.887" W., NAD 1983–86), SR-28 (latitude 32° 06' 08.521" N., longitude 081° 00' 15.498" W., NAD 1983–86), and SR-27 (latitude 32° 06' 01.565" N., longitude 080° 59' 58.406" W., NAD 1983–86), Upper Flats Range through points SR-26 (latitude 32° 05' 41.698" N., longitude 080° 59' 31.968" W., NAD 1983–86) and SR-25 (latitude 32° 05' 02.819" N., longitude 080° 59' 12.644" W., NAD 1983–86), Lower Flats Range through points SR-24 (latitude 32° 04' 46.375" N., longitude 080° 59' 00.631" W., NAD 1983–86), SR-23 (latitude 32° 04' 40.209" N., longitude 080° 58' 49.947" W., NAD 1983–86), SR-22 (latitude 32° 04' 28.679" N., longitude 080° 58' 18.895" W., NAD 1983–86), and SR-21 (latitude 32° 04' 22.274" N., longitude 080° 57' 34.449" W. , NAD 1983–86), Long Island Crossing Range through points SR-20 (latitude 32° 04' 13.042" N., longitude 080° 57' 14.511" W., NAD 1983–86), and SR-19 (latitude 32° 02' 30.984" N., longitude 080° 55' 30.308" W., NAD 1983–86) and New Channel Range following the northern boundary of the Rehandling Basin and the northern boundary of the Oyster Bed Island Turning Basin back to the northern edge of the main navigational channel, thence through points SR-17 (latitude 32° 02' 07.661" N., longitude 080° 53' 39.379" W., NAD 1983–86) and SR-16 (latitude 32° 02' 07.533" N., longitude 080° 53' 31.663" W., NAD 1983–86), to a point at latitude 32° 02' 08" N., longitude 080° 53' 25" W., NAD 1983–86 (now marked by Navigational Buoy "24") near the eastern end of Oyster Bed Island;

Proceeding thence from a point at latitude 32° 02' 08" N., longitude 080° 53' 25" W., NAD 1983–86 (now marked by Navigational Buoy R "24") on a true azimuth of 0° 0' 0" (true north) to the mean low low-water line of Oyster Bed Island; thence easterly along the said mean low low-water line of Oyster Bed Island to the point at which the said mean low low-water line of Oyster Bed Island intersects the Oyster Bed Island Training Wall;

Proceeding thence easterly along the mean low low-water line of the southern edge of the Oyster Bed Island Training Wall to its eastern end; thence continuing the same straight line to its intersection with the Jones Island Range line;

Proceeding thence southeasterly along the Jones Island Range line until reaching the northern boundary of the main navigational channel as depicted on the Channel Chart;

Proceeding thence southeasterly along the northern boundary of the main navigational channel as depicted on the Channel Chart, via Jones Island Range and Bloody Point Range, to a point at latitude 31° 59' 16.700" N. , longitude 080° 46' 02.500" W., NAD 1983–86 (now marked by Navigational Buoy "6"); and finally,

Proceeding from a point at latitude 31° 59' 16.700" N., longitude 080° 46' 02.500" W., NAD 1983–86 (now marked by Navigational Buoy "6") extending southeasterly to the federal-state boundary on a true azimuth of 104 degrees (bearing of S76°E), which describes the line being at right angles to the baseline from the southernmost point of Hilton Head Island and the northernmost point of Tybee Island, drawn by the Baseline Committee in 1970.

Should the need for further delimitation arise, the boundary shall further extend southeasterly on above-described true azimuth of 104 degrees (bearing of S76°E).

Provided, further, that nothing in this section shall in any way be considered to govern or affect in any way the division between the states of the remaining assimilative capacity that is, the capacity to receive wastewater and other discharges without violating water quality standards, of the portion of the Savannah River described in this section.

The importance that Georgia and South Carolina give to their descriptions of the Savannah River as the boundary between the two states is reflected in the detail they use to delineate the boundary. Evidently both states want to be precise to avoid future misunderstandings.

In the past, Georgia and South Carolina have had numerous lawsuits in the U.S. Supreme Court over their mutual boundary. Beginning with the vague boundary description in 1732 as part of the Trustee's Charter and continuing until the late 20th century, the two states have had ongoing disputes involving either navigation, commerce or ownership of various islands in the river. The U.S.

Army Corps of Engineers has also dramatically changed the direction and orientation of the river, resulting in claims by both states based on accretion or erosion.

Commerce on the river has always been of great concern to both Georgia and South Carolina. As early as 1736, only three years after Georgia was settled, South Carolina made a complaint to England's Board of Trade when Georgia began requiring South Carolina vessels sailing up the Savannah River have a special permit to trade with the natives. South Carolina also objected to Georgians, who did not allow "rum or other spirits" in the colony, boarding South Carolina vessels to search for and seize any rum or other alcoholic beverages on board.

Two questions which the Board of Trade had to answer were first, whether any colonial act could grant itself exclusive trade with the Native Americans, and second, whether Georgia could exclude all persons without a Georgia license from trading any product with the Native Americans. The Board of Trade opined that no colony could grant itself exclusive trading and that Georgia could require all traders to have a Georgia license to trade with the local tribes. The Board of Trade further ruled that Georgia could not arbitrarily refuse anyone a license and specifically required Georgia to give a license to any South Carolina merchant who requested one. Even though the ruling was against Georgia, it was largely ignored by both colonies during the Trustees' period in Georgia. Of more importance, the ruling undermined the Trustees' authority in the colony for the first time.

The first lawsuit heard by the U.S. Supreme Court between Georgia and South Carolina was South Carolina v. Georgia (93 U.S. 4) in 1876.

The Savannah River, which flows past the city of Savannah, is divided into two channels, each with a length of about six miles. In 1876 the U.S. Congress had appropriated $70,000 for the improvement of the harbor at Savannah, and the Corps of Engineers of the U.S. Army constructed a dam above the city of Savannah to divert water passing through the north channel Back River into the Front River channel, which provided the city with 15 feet of depth at low tide. South Carolina filed an injunction restraining Georgia and the Corps of Engineers from performing any work on the river which would result in the obstruction of the state's ability to navigate the river. South Carolina told the Court that while this diversion may have helped with the navigation of one channel of the river, it interfered with and obstructed the other channel.

South Carolina cited Article 2 of the 1787 Beaufort Convention.

> The navigation of the River Savannah, at and from the bar and mouth, along the northeast side of Cockspur Island, and up the direct course of the main northern channel, along the northern side of Hutchinson's Island, opposite the town of Savannah, to the upper end of the said island, and from thence up the bed or principal stream of the said river to the confluence of the Rivers Tugoloo and Keowee, and from the confluence up the channel of the most northern stream of Tugoloo River to its source, and back again by the same channel to the Atlantic Ocean, is hereby declared to be henceforth equally free to the citizens of both States, and exempt from all duties, tolls, hindrance, interruption, or molestation whatsoever attempted to be enforced by one State on the citizens of the other, and all the rest of the River Savannah to the southward of the foregoing description is acknowledged to be the exclusive right of the State of Georgia.

It relied on the words that navigation on the Savannah River was equal to citizens of both states, exempt from molestation of one state on the other.

The court denied South Carolina's request for the injunction and dismissed the case. The justices' reasoning was that when both Georgia and South Carolina became part of the United States in 1889, they adopted the U.S. Constitution which delegated to the federal government the right to "regulate commerce with foreign nations, and among the several states." The court went on to say, "Interstate commerce extends to the control of navigable rivers between states and Congress has the power to delegate changes to the river to provide greater commerce on the river. The fact that another channel of the river is negatively affected is of less value than that received from the increased commerce."

The next lawsuit was Georgia v. South Carolina (257 U.S. 516,

1922) filed by Georgia's attorney general in 1917. Both states agreed that the Savannah River was the boundary line between the two, but the question in this case concerned the exact location of the boundary within the river itself. The decision of case would affect the taxation of dams and hydroelectric plants either under construction or scheduled for future construction on the river. In addition, both states asked the court to decide who owned the islands in the Savannah River and which state owned any islands that were located in the Chattooga River, a tributary of the Tugaloo River but not named in the 1787 Beaufort Convention.

The Court referred to Article 1 of the Beaufort Convention which defines the boundary between the two states as:

> The most northern branch or stream of the River Savannah from the sea or mouth of such stream to the fork or confluence of the Rivers then called Tugaloo and Keowee; and from thence the most northern branch or stream of said River Tugaloo, till it intersects the northern boundary line of South Carolina, if the said branch or stream of Tugaloo extends so far north, reserving all the islands in the said Rivers Savannah and Tugaloo, to Georgia; but if the head, spring, or source of any branch or stream of the said river Tugaloo does not extend to the north boundary line of South Carolina, then a west course to the Mississippi, to be drawn from the head, spring, or source of the said branch or stream of Tugaloo river, which extends to the highest northern latitude, shall forever thereafter form the separation, limit, and boundary between the States of South Carolina and Georgia.

The court concluded that because a branch or stream of the Tugaloo River does extend as far north as the northern boundary line of South Carolina (35° north latitude), all the islands in the Savannah and Tugaloo Rivers belonged to Georgia. Because subsequent exploration revealed that the Tugaloo River is formed when the Chattooga River comes from the north to join the Tallulah River, any islands in the Chattooga River also belong to Georgia. The fact that the Chattooga River is not specifically mentioned in Article 1 does not change the intent. The court lastly discussed where the boundary would be whether the river contained or did not contain islands. The decision reads:

> Thus we conclude: (1) Where there are no islands in the boundary rivers the location of the line between the two states is on the water midway between the main banks of the river when the water is at ordinary stage; (2) where there are islands the line is midway between the island bank and the South Carolina shore when the water is at ordinary stage; and (3) that islands in the Chattooga river are reserved to Georgia as completely as are those in the Savannah or Tugaloo rivers.

The resolution of the 1922 case seemed straightforward and indisputable. Unfortunately, the direction and the topography of a river can change, and the two states would return to the courts for a decision about which owned an island in the river.

Savannah is a major port, and huge tankers make their way upstream every day to be offloaded. The 30-mile distance of the Savannah River from its mouth at the Atlantic Ocean to the port of Savannah is convoluted with channels, crossing tributaries, islands and marshes. To keep the river navigable and safe, the Corps of Engineers frequently dredges the river and deposits the silt on small islands or spits of land near the South Carolina side of the river. Over the years, this constant dredging has significantly changed the contour of this portion of the river.

The 1955 case of the United States of America v. 450 Acres of Land, More or Less, Known as Barnwell Island (220 F.2d 353, cert. denied; 350 U.S. 826, 1955) was brought in Federal District Court in Georgia in 1955. The United States wanted to acquire, by condemnation, a perpetual and assignable easement to enter upon, occupy and use certain lands known as Barnwell Island for the deposit of spoil or other matters excavated in the construction, improvement and maintenance of Savannah Harbor. E.B. Pinckney appeared at the condemnation hearing and told the court that because he owned Barnwell Island and had been paying taxes on the island to South Carolina for 10 years, the hearing should be dismissed because the island was not in Georgia. Pinckney wanted the condemnation hearing brought in South Carolina so that a proper value could be place on the property.

Considering the legal doctrines of prescription and acquiescence, the District Court dismissed the United States' condemnation case.

As discussed earlier, prescription is the open, continuous and exclusive use of property, while acquiescence is the failure to properly assert one's rights. Pinckney had openly used the property for 10 years, and Georgia never disputed the sale nor attempted to collect any property taxes.

Furthermore, Barnwell Island had been attached for many years to the South Carolina shore because of changes in the river contour, and Georgia never asserted its claim to the property. As part of their policy to periodically scrutinize the Atlantic coastline, the U.S. Geological Survey redrew topographical maps of the 30-mile lower part of the Savannah River in 1955. The map showed small islands in the north channel which used to be in the middle of the channel were now attached to the South Carolina shore of the river because of the dredging and silt deposits. For the first time, the map showed a dotted line purporting to be the boundary between Georgia and South Carolina. The dotted line went up the middle of the north channel of the river — the only navigable channel — and specifically placed Barnwell Island on the South Carolina side.

The United States appealed the District Court's decision to the U.S. Court of Appeals. This court reviewed the evidence and came to an opposite conclusion.

> The boundary line between Georgia and South Carolina is not in dispute as between these sovereigns. Neither the State of South Carolina nor any official purporting to act for South Carolina has intervened, become a party, or made any appearance in this proceeding and even if Georgia could lose territory by acquiescence in an assertion of jurisdiction by South Carolina, this record contains nothing of substance to show either acquiescence or assertion. It is true that Georgia did not tax the land but there is nothing to show that this was because the State did not believe that the land was within its jurisdiction. When South Carolina sanctioned the compact [Beaufort Convention] it admitted in the most solemn form that the land in controversy was not within her jurisdiction and it is certainly reasonable to assume that Georgia believed it owned the land. Consequently, there was no occasion for it or any of its political subdivisions to impose a tax. On the other hand, South Carolina has not claimed jurisdiction over the land

in controversy. It has not been shown that the taxes, for which the land was sold to the Forfeited Land Commission of Beaufort County, and by it to Pinckney, were imposed by the State or applied to the benefit of the State. ... There is no proof here that Georgia had knowledge of, much less that it acquiesced in, these isolated activities and it is absurd to say that a state has relinquished a right of which it is not aware. Furthermore, there is nothing here to connect the state of South Carolina with these transactions and she does not and cannot claim that she acquired any right by virtue of these activities.

The Appeals Court concluded that Barnwell Island was in the state of Georgia, that the condemnation proceeding was properly brought in the federal District Court, and that it had been error for the District Court to dismiss it. The Appeals Court directed the parties to return to the District Court for the case to be heard.

It is unknown if Georgia or South Carolina even noticed the boundary line in the 1955 map, but whether they did or not, since no one lived in the area and the U.S. Geological Survey had no power to decide or alter boundary line, neither state voiced a complaint. The topographical maps were again redrawn 16 years later using aerial photographs and reviewing historical maps of the region from the time of the Beaufort Convention. The resulting map revised the boundary line and returned some of the attached islands to Georgia.

South Carolina finally sprang into action in 1977. Noticing that the 1971 map had changed the boundary line from the 1955 map, South Carolina's legislature passed a resolution, "To Declare that the United States Geological Survey Should Immediately Begin Consultation with the Authorized Representatives of Both South Carolina and Georgia in Order to Correct the Erroneous Delineation of the Boundary Line Between South Carolina and Georgia." Both states agreed to discuss the situation. A conference was held with a presentation by an agent of the U.S. Geological Survey who stated, "The United States Geological Survey does not establish boundaries nor does it adjudicate boundary disputes."

Attempts were made to placate South Carolina by redrawing a

map without boundary lines depicted, but two other problems contributed to heating up the controversy: commercial fishing and offshore oil.

Commercial fishing played an important economic role for both states, which had different laws regarding the seasons, locations and times, and the two states were also meeting to discuss these issues. One incident in particular brought the fishing problem to a head. In July 1977, a shrimp boat captain was arrested by a conservation ranger from the Georgia Department of Natural Resources and charged with illegally fishing in Georgia waters without a commercial saltwater fishing license. Immediately after his arrest, the captain was told to follow the ranger back to Savannah, but he failed to follow instructions and raced to Hilton Head instead. Georgia requested that South Carolina extradite the captain to Georgia for trial; South Carolina refused saying that the arrest was illegal since the fishing was done in South Carolina waters.

The waters offshore became a point of contention. During the early 1970s, a Middle Eastern embargo on oil escalated gasoline prices and brought the United States to a realization of its serious dependency on foreign oil and its need for domestic oil sources. As a result, the U.S. Congress passed the United States Coastal Zone Management Act of 1972 granting millions of dollars of federal aid to coastal states for energy exploration, based on activity on the outer continental shelf "adjacent" to that state. The adjacency was to be determined, in the absence of boundaries already established on the basis of interstate agreements or court decisions, by lateral boundaries drawn in the ocean seaward from the coastal state extending for three nautical miles. Georgia and South Carolina disagreed on the exact location of the mouth of the Savannah River that would serve as the starting point for the "lateral seaward boundary" defining the border between them in the ocean.

After months of negotiations, it became clear that only litigation could resolve these disputes. In August 1977, Georgia filed suit in the U.S. Supreme Court asking the court to decide the exact location of the lateral seaward boundary and whether man-made changes in the river could alter the 1787 Beaufort Convention regarding the ownership of Barnwell, Jones and Oyster Bed islands as well as some spoil islands which resulted from dredging of navigational channels.

The Supreme Court agreed to hear Georgia v. South Carolina (497 U.S. 376, 1990) and appointed Senior Federal Judge Walter E. Hoffman of the U.S. District Court for the Eastern District of Virginia as a special master to act on their behalf and conduct hearings where both states could present their arguments. Such hearings include oral arguments, testimony of expert witnesses who are subjected to direct and cross examination, and presentation of maps and historical documents. After hearing all the evidence, the special master then renders his report to the court and the parties, at which time each side is allowed to file a written exception to the report with the court.

Judge Hoffman tendered his first report to the court in 1986, in which he discussed and opined on all of the issues except for the lateral seaward boundary, which was addressed in his final report tendered to the court three years later. In January 1990 both states submitted their exceptions to the two reports, and that July, 13 years after the initial filing of the lawsuit, the Supreme Court rendered its decision.

The court's 40-page opinion is extremely detailed with historical information about Georgia and the Savannah River, technical data, review of case law from other boundary disputes in the United States, and a well-reasoned discussion of its conclusions. For the most part, they concurred with Hoffman.

1. The Barnwell Islands are in South Carolina, which had established sovereignty by prescription and acquiescence.
2. Islands emerging in the river after the 1787 Beaufort Convention would not affect the boundary line between the states. The intent of the convention was to establish a permanent boundary between the two states and not one that would shift because of nature or humankind's interference. Any new island belongs to the state which owned the river bed from which it was formed.
3. Oyster Bed Island belongs to South Carolina.
4. The Denwill and Horseshoe shoals belong to Georgia since these were formed by dredging and other processes used by the

Army Corps of Engineers to improve the river's navigation channel.
5. The mouth of the Savannah River is bounded on the southern side by Tybee Island and on the northern side by an underwater shoal.
6. The lateral seaward boundary extends the boundary at the Savannah River's mouth until it intersects a line drawn from Tybee Island's most northern point to Hilton Head Island's most southern point, continuing perpendicularly from that line eastward out to sea.
7. Jones Island belongs to South Carolina as it has never been in the Savannah River.

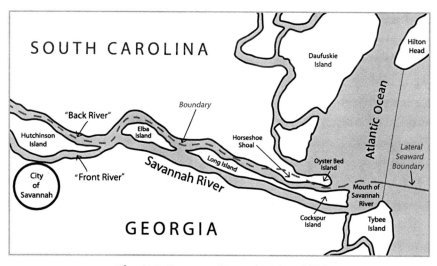

After U.S. Supreme Court decision, 1990

The only issue on which the court disagreed with Hoffman was that of drawing a right-angle boundary around small islands on the South Carolina side of the river rather than the northern-most shore line of the island to determine the boundary. The court agreed with Georgia's assessment that the boundary "is to be marked by the use of a point which is tri-equidistant from the South Carolina shore, the island shore, and the Georgia shore."

The court's opinion also required Georgia and South Carolina to survey the lower portion of the Savannah River to comply with the court's findings. Neither state was not in a hurry to do so. Almost a decade after the court's decision and 22 years after the suit was brought, the states agreed on their mutual boundary and submitted their survey to the court in 1999, which sent it to the U.S. Congress for approval. Careful reading of each state's boundary description in their respective codes reveals that both agree on the boundary line, although South Carolina has added more specific latitude and longitude information.

Very little property changed hands as a result of the 1977 lawsuit. Georgia gained a few small spoil islands, while South Carolina gained the land attached to its shore. Commercial fishing limits were better defined, as was the mouth of the river and the lateral seaward boundary.

In the future, using an underwater shoal as a boundary line for the mouth of the river may become problematic as nature changes the shifting sands. The constant dredging of the river to provide safe navigation will also continue to accumulate silt, and the resultant spoil islands which could serve a recreation and conservation purpose may be of interest to investors for commercial development as well. It seems highly probable that the two states will see each other again in a courtroom.

Conclusion

For the most part, conflicts over ownership or use of the land and rivers have been examined and resolved, and, with two notable exceptions, the boundaries of the state have been agreed upon by Georgia and its neighbors.

The boundaries of Georgia are more than mere lines on a map. From the description in its original charter in 1732 to the familiar shape we recognize today, many events and people — both well-known and relatively unheard of — have left their mark on the land. The boundary lines tell their story. Georgia's boundaries are a meeting of history and geography.

Important Dates V

1851	Howard v. Ingersoll (Georgia/Alabama)
1854	Florida v. Georgia
1859	Alabama v. Georgia
1876	South Carolina v. Georgia
1887	Coffee v. Groover (Georgia/Florida)
1889	Tennessee agrees to joint survey of boundary
1897	Georgia passes resolution regarding Tennessee boundary
1898	Wimbish Report issued (Georgia/Tennessee)
1905	Tennessee agrees to joint survey of boundary
1915	Georgia and Tennessee governors agree to meet regarding boundary
1916	Georgia passes resolution regarding Tennessee boundary
1922	Georgia v. South Carolina
	Georgia passes resolution regarding Tennessee boundary
1930	Florida Gravel Company v. Capital City Gravel Company
1941	Georgia appoints standing committee to meet with Tennessee to resolve boundary
1947	Tennessee governor meets with Georgia delegation regarding boundary
1955	United States of America v. 450 Acres of Land, More or Less, known as Barnwell Island (Georgia/South Carolina)
1971	Georgia governor authorized to create commission to resolve Georgia/Tennessee boundary
1990	Georgia v. South Carolina
2008	Georgia passes resolution to correct northern boundary with Tennessee and North Carolina
	Tennessee passes resolution declining to participate in any boundary negotiation

Epilogue

So there you have it. Quite a story, isn't it? As the saying goes, "Truth is stranger than fiction."

I have been looking at charts and maps all my adult life and have never given any state's boundaries a second thought. Although I grew up in Florida and spent the last 50 years in Georgia, in January 2006, while driving from Savannah to Jacksonville and crossing over the St. Marys River, I noticed for the first time that the river was the boundary between the two states.

I unfolded the map and looked carefully at all of Georgia's boundaries. My mind went into high gear. Who was the first to come up with the boundaries? How were they decided? Who were the people who walked them? What instruments did they use to measure them? Were the boundaries ever changed? On and on and on.

The first thing I did was to go online to see if anyone else had written such a book. When I didn't find anything, I became emboldened to be the first person to do so. I can remember thinking that I had finally thought of something that no one else in the world had ever thought of — maybe I'd discovered my golden parachute.

It wasn't until months later that I found numerous magazine articles or children's books about a particular state's boundaries, biographies about prominent surveyors, and surveying magazines with historical articles about boundaries. Halfway through my research, I came across Marion Hemperley and Ed Jackson's *Georgia's Boundaries: The Shaping of a State*. Although out of print, this excellent book became the benchmark (fortuitous use of a surveyor's term) for me to meet.

The journey I've taken in writing this book has given me simultaneous pleasure and frustration, and my life is much richer for the experience. My appreciation for the evolution of Georgia's boundaries grew and the story became more interesting when the personalities came to life. For the past three years, I have been living in the 17th and 18th centuries. The amazing people from that era who accomplished so much opened up new avenues of further exploration for me.

- Richard Hakluyt, the clergyman who galvanized England and his queen to explore the New World.
- James Oglethorpe, whose controlling role in the colony was largely responsible for its failure, while his astounding defeat of the Spanish at the Battle of Bloody Marsh in 1742 clearly allowed England to gain control of the North American continent. Oglethorpe spent only 10 years in Georgia but carried on a long and interesting life in England for the next half-century.
- Thomas Stephens, a founding member of the Georgia Malcontents and the dissident son of the president of Georgia, who wrote articles and appeared in Parliament to tell his version of conditions in the new colony.
- James Wright, the third royal governor of Georgia, best remembered for his excellent administration of the colony and his sensible dealing with the Native Americans.
- Thomas Paine, the uneducated, self-styled philosopher, whose writings continue to stir the hearts and minds of people everywhere. Who cannot hear "These are the times that try men's souls" and not be moved?
- I can't say enough about Andrew Ellicott, friend of Washington, Jefferson and Franklin and a brilliant professional so important to the history of the United States — yet unknown to most Americans. Reading his four-year journal of surveying the United States/Spain border revealed the depth of his knowledge about astronomy, mathematics, botany, and social, political and economic issues and piqued my curiosity about this man.
- Finally, there is a villain in every story. That dubious honor here belongs to James Camak, the incompetent mathematician and surveyor whose faulty measurement of Georgia's northern boundary continues to haunt the state.

My goal in writing this book was to tell a unique story. Nothing unusual about that for an author, and I trust that I have done that successfully. Perhaps this book will be the impetus for another author to tell his or her story.

Monna J. Morton

About the Author

William J. Morton is a board-certified urological surgeon who practiced in Atlanta for 30 years. He is a graduate of the University of Miami School of Medicine and did his post-graduate urology training at the Emory-Grady Residency Program in Atlanta. He served as a medical officer in the U.S. Air Force, sat on numerous hospital committees, and has published over two dozen articles in medical journals.

Bill received his Doctor of Jurisprudence degree in 1985 and has been a member of the State Bar of Georgia since that time. He has taught courses, written papers and made presentations on medico-legal topics throughout the United States. He is also a part-time Magistrate Court judge in Fulton County, Ga.

His wide interests include history, astronomy, ornithology and photography, and he holds private pilot and U.S. Coast Guard captain licenses. He has already started doing research on his next book, a biography of Andrew Ellicott.

Bill and his wife, Monna, both passionate about fly fishing, live in Atlanta and in Ennis, Montana.

www.wjmortonmdjd.com

Acknowledgements

When writing a book — particularly a non-fiction book where one has little knowledge about the topic — authors must rely on many different types of resources. Gathering the information for this book required various investigative techniques and was equally as difficult as the writing of the actual text. I spoke to dozens of complete strangers, probing them for information, and never was I rejected or treated rudely. All went out of their way to guide and assist me and would follow up with suggestions about finding the answers I sought.

Three individuals have been most important to me for the completion of this project. Amy Benson Brown, Ph.D, director of the author development program of the Center for Faculty Development and Excellence at Emory University, was the first to read the entire manuscript and provided invaluable constructive criticisms, all of which I adopted. Kate Siegel of Kartouche in Atlanta, my editor and book designer, and John C. Nelson, an Atlanta-based illustrator, took the sow's ear of my raw data and turned it into a silk purse. The finished product is more than I imagined.

A special thanks goes to Steve Engerrand, assistant director of archival services at the Georgia Archives, for permission to photograph and publish the image of the Original Trustees Seal of Georgia, 1732, and to reproduce the "plat for survey, 887 acre bounty grant to William Few, 1784–85" from Record Group 3 (Records of the Surveyor General) in the cover artwork.

The following institutions, historical societies, libraries, museums, and internet sites aided my research:
Alabama Historical Society
American Museum of History, Smithsonian Library

American Philosophical Society
Avalon Project, Yale University
Capital Gallery Archives, Washington, D.C.
Carl Vinson Institute of Government, University of Georgia
Catholic Online Encyclopedia
Dictionary of Canadian Biography Online
Emory University Law Library
Encyclopedia Britannica Online
The Field Museum, Chicago
Florida Historical Society
Fulton County Courthouse Law Library
Fulton County Public Library
Georgia Archives
Georgia Historical Society
Library of Congress – Geography and Map Division
Library of Congress – Manuscript Division
Louisiana Historical Society
MSN Encarta
New Georgia Encyclopedia
North Carolina Historical Society
St. Marys Chamber of Commerce
South Carolina Historical Society
Tennessee Historical Society
U.S. Board on Geographic Names
Wikipedia

Finally, my appreciation extends to the following for their personal communications:
Ed Jackson, co-author of *Georgia's Boundaries: The Shaping of a State*
David Bederman, professor of law, Emory University
Robert Shannon, environmental attorney, Hall, Booth, Smith and Slover, P.C., Atlanta
Andrew Lannen, Ph.D., assistant professor, Louisiana State University
Gregory Spies, professional land surveyor, Mobile, Ala.
Tammy Peters, supervisory archivist, Smithsonian Institution Archives
Jane Basnight, reference librarian, Legislative Library of the North

Carolina General Assembly

John Briscoe, Esq., Briscoe, Ivester and Bazel, LLP, San Francisco, special counsel to state of Georgia in Georgia v. South Carolina (475 U.S. 1115, 1986)

Brad Carver and Bobby Shannon, attorneys at law, Hall, Booth, Smith and Sullivan Law Offices, Atlanta

Lisa A. Macklin, J.D., coordinator, Intellectual Property Rights Office of Emory University Libraries

I would be dismayed to find that I left some person or institution out. If so, unidentified person or place, please forgive me.

Bibliography

Akerman, James R., and Robert W. Karrow Jr., eds. *Maps: Finding our Place in the World.* The University of Chicago Press, Chicago: 2007. (ISBN 978-0-226-01075-5)

Alabama, State of. *Code of Alabama*, §41-2-2: 1975. Digital text Alabama Legislature, 1997. <http://www.legislature.state.al.us/codeofalabama/1975/41%2D2%2D2.htm>

Alabama, The State of, v. The State of Georgia. 64 U.S. 505, 23 How. 505, 16 L.Ed. 556, 1859. Digital text Open Jurist. <http://openjurist.org/64/us/505/the-state-of-alabama-v-the-state-of-georgia>

Allen, Alexander A. "Allen's Journal: A Trip Along the Georgia–Florida Boundary, June 14–July 22, 1854," *Occasional Paper from South Georgia*, no. 6, eds. Christy T. Trowell and Frances R. Trowell. South Georgia College, Douglas, Ga.: 1984.

Arthur, John Preston. *Western North Carolina, A History from 1730–1913.* Edward Buncombe Chapter of the Daughters of the American Revolution, Asheville, N.C.: 1914. Reprint The Overmont Press, Johnson City, Tenn.: 1996. (ISBN 1-57072-062-2)

Arthur, T.S. and W.H. Carpenter. *The History of Georgia: From Its Earliest settlement to the Present Time.* Lippincott & Co., Philadelphia: 1853. Reprint Kessinger Publishing's Legacy Reprints, Whitefish, Mont.: 1948.

Articles of Agreement and Cession, 1802. In *Georgia Land Surveying History and Law*, Ferris W. Cadle.

Bailey, Liberty H. *Sketch of the Evolution of Our Native Fruits.* MacMillan Company, London: 1898. Digital reproduction Harvard University, Boston: June 2007.

Banning, Lance. "From Confederation to Constitution: The Revolutionary Context of the Great Convention," *This Constitution: A Bicentennial Chronicle.* Project '87, American Political Science Association and American Historical Association, Washington, D.C.: 1985. (http://www.apsanet.org/imgtest/RevolutionConfederation.pdf)

Barnes, A. Roger. *The State of North Carolina: A History of the Boundaries Surveyed.* Map. North Carolina Geodetic Survey, Raleigh. <http://www.esri.com/mapmuseum/mapbook_gallery/state1/nc1.html>

The Beaufort Convention, 1787. In Georgia v. South Carolina.

Bedini, Silvio. "Andrew Ellicott, Surveyor of the Wilderness," *Quarterly Journal of the American Congress on Surveying and Mapping,* vol. 26, no. 2 (June 1976): p. 113–135.

Bedini, Silvio. "Early American Surveyors: Mapping the Wilderness," *Professional Surveyor Magazine,* Mar. 1982: p. 24–28.

Bedini, Silvio A. "Gustavus John Orr (1819–1887): Georgia Educator and Surveyor, Part I," History Corner, *Professional Surveyor Magazine,* vol. 23, no. 5 (May 2003). <http://www.profsurv.com/magazine/article.aspx?i=1077>

Bedini, Silvio A. "Gustavus John Orr (1819–1887): Georgia Educator and Surveyor, Part II," History Corner, *Professional Surveyor Magazine,* vol. 23, no. 6 (June 2003). <http://www.profsurv.com/magazine/article.aspx?i=1088>

Bedini, Silvio. *The Jefferson Stone: Demarcation of the First Meridian of the United States.* Professional Surveyors Publishing Co., Frederick, Md.: 1999. (ISBN 0-9665120-1-4)

Bedini, Silvio. *The Life of Benjamin Banneker: The First African-American Man of Science,* 2d ed. Professional Surveyors Publishing Co., Frederick, Md.: 1999. (ISBN 0-938420-63-1)

Bedini, Silvio. *With Compass and Chain: Early American Surveyors and Their Instruments*. Professional Surveyors Publishing Co., Frederick, Md.: 2001. (ISBN 0-9665120-0-6)

Boorstin, Daniel J. *The Discoverers*. Penguin Books:1983. (ISBN 978-03947-2625-0)

Bowen, Catherine Drinker. *Miracle at Philadelphia: The Story of the Constitutional Convention, May to September, 1787*. Atlantic Monthly Press, New York: 1966. (ISBN 978-03161-0398-5)

Bryant, Pat. *Georgia Counties: Their Changing Boundaries*. State Printing Office, Atlanta: 1977

Bryant, Pat. *Georgia State Boundary Documents: Florida–Georgia Boundary, a Calendar*. State Printing Office, Atlanta: 1978

Cadle, Ferris W. *Georgia Land Surveying History and Law*. University of Georgia Press, Athens, Ga.: 1991 (ISBN 0-8203-1257-6)

Camak, James. Report to Georgia, 1826. In "The Georgia-Tennessee Boundary Line," E. Merton Coulter, and *North Carolina Boundary Disputes Involving Her Southern Line*, Marvin L. Skaggs.

Candler, A.D., and A.E. Clement, eds. *Georgia: Comprising Sketches of Counties, Towns, Events, Institutions and Persons arranged in Cyclopedic Form, Vol. I*. State Historical Association, Atlanta: 1906.

Carter, Clarence E., ed. *The Territorial Papers of the United States: Territory of the United States South of the River Ohio, 1790–1796*, vol. 4. Government Printing Office, Washington, D.C.: 1936.

Charter of Carolina, 1663. Digital text "Charter of Carolina — March 24, 1663," *Avalon Project*, 2008. Lillian Goldman Law Library, Yale Law School, New Haven, Conn. <http://avalon.law.yale.edu/17th_century/nc01.asp>

Charter of Georgia, 1732. Digital text "Charter of Georgia: 1732," *Avalon Project*, 2008. Lillian Goldman Law Library, Yale Law School, New Haven, Conn. <http://avalon.law.yale.edu/18th_century/ga01.asp>

Columbus, Christopher. *The Log of Christopher Columbus*, trans. Robert Fuson. International Marine Publishing Co., Camden, Me.: 1987. (ISBN 0-87742-951-0)

Coffee v. Groover. 123 U.S. 1, 1887. Digital text Open Jurist. <http://www.openjurist.org/123/us/1/coffee-v-groover>

Coulter, E. Merton. "The Georgia-Tennessee Boundary Line," *The Georgia Historical Quarterly*, vol. 35, no. 4 (Sept. 1951).

De Brahm, William Gerard. *De Brahm's Report of the General Survey in the Southern District of North America.*, ed. Louis de Vorsey, Jr. 1799. Reprint University of South Carolina Press, Columbia: 1987. (ISBN 0-87249-229-X)

De Vorsey, Louis, Jr. *The Georgia-South Carolina Boundary: A Problem in Historical Geography*. University of Georgia Press, Athens, Ga.: 1982. (ISBN 0-8203-0591-X)

Doub, William C. *A History of the United States*. Doub & Company, San Francisco: 1906.

Duncan, David Ewing. *Calendar: Humanity's Epic Struggle to Determine a True and Accurate Year*. Avon Books, New York: 1998. (ISBN 978-038079-324-2)

Ehrenberg, Ralph E. *Mapping the World: An Illustrated History of Cartography*. National Geographic, Washington, D.C.: 2001. (ISBN 978-079226-525-2)

Ellicott, Andrew. *The Journal of Andrew Ellicott for Determining the Boundary Between the United States and the Possessions of His Catholic Majesty in America containing occasional remarks on the Situation, Soil, Rivers, Natural Productions, and Diseases of the Different Countries on the Ohio, Mississippi, and Gulf of Mexico with Six Maps to which is added an Appendix Containing all the Astronomical Observations Made Use of for Determining the Boundary*. Budd and Bartram for Thomas Dobson, Philadelphia: 1803. Reprint Arno Press, New York: 1980. (ISBN 0-4051-2541-0)

Ellicott, Andrew. *Several Methods by which Meridional Lines are Found with Ease and Accuracy.* Thomas Dobson, Philadelphia: 1796. Reprint American Philosophical Society, Philadelphia.

Ferling, John. *Almost A Miracle: The American Victory in the War of Independence.* Oxford University Press: 2007. (ISBN 978-0195-18121-0)

Fernández-Armesto, Felipe. *Columbus.* Oxford University Press: 1991. (ISBN 0-19-215898-8)

Fiske, John. *The Critical Period of American History, 1783–1789.* Houghton Mifflin Company, Boston: 1888.

Florida, State of. *The 2009 Florida Statutes,* §6.09: 2009. Digital text Florida Legislature, 1995–2009. <http://www.leg.state.fl.us/STATUTES/index.cfm?App_mode=Display_Statute&Search_String=&URL=Ch0006/SEC09.HTM&Title=-%3E2009-%3ECh0006-%3ESection%2009#0006.09>

Florida v. Georgia. 58 U.S. 478, 1854. Digital text Justia & Oyez & Forms WorkFlow. <http://supreme.justia.com/us/58/478/case.html>

Florida Gravel Co. v. Capital City Sand & Gravel Co. 170 Ga. 855, 1930. Digital text *Casemaker,* State Bar of Georgia.

Georgia, State of. "Boundary Between Georgia and Florida," §15-105 Amended, *Acts and Resolutions of the General Assembly of the State of Georgia, 1969,* vol. 1, p. 675.

Georgia, State of. *The Colonial Records of the State of Georgia,* vol. 9. Franklin-Turner Company, Atlanta: 1907.

Georgia, State of. *Georgia Constitution of 1798.* Digital text Carl Vinson Institute of Government, University of Georgia. <http://georgiainfo.galileo.usg.edu/con1798.htm>

Georgia, State of. "'Georgia Day,' February 12, Observance by Public Schools," *Georgia Laws,* 1909. p. 190. Digital text Carl Vinson Institute of Government, University of Georgia. <http://georgiainfo.galileo.usg.edu/georgiaday.htm>

Georgia, State of. *Official Code of Georgia Annotated*: 2009. Digital text LexisNexis, 2009. <http://www.lexis-nexis.com/hottopics/gacode/default.asp>

Georgia, State of. Resolution: Dec. 21, 1897. *Georgia Laws*, 1897. p. 595.

Georgia, State of. Resolution: Aug. 17, 1906. *Georgia Laws*, 1906. p. 1160.

Georgia, State of. Resolution: Aug. 16, 1922. *Georgia Laws*, 1922. p. 1139.

Georgia, State of. Resoultion: Mar. 6, 1941. *Georgia Laws*, 1941. p. 1850.

Georgia, State of. Senate Resolution 822: 2008. Digital text Georgia General Assembly. <http://www.legis.state.ga.us/legis/2007_08/fulltext/sr822.htm>

Georgia v. South Carolina. 497 U.S. 376, 1990. Digital text Open Jurist. <http://openjurist.org/497/us/376/georgia-v-south-carolina>

Goetchius, Henry R. "Great Seals of Georgia: Their Origin and History," *The Georgia Historical Quarterly*, vol. 1, no. 3 (Sept. 1917): p. 253–260. Digital reproduction Google, 2008. <http://books.google.com/books?id=HrwKAAAAIAAJ&pg=RA1-PA253&lpg=RA1-PA253&dq=Great+Seals+of+Georgia%3B+Their+Origin+and+History&source=bl&ots=ZNFAOER09S&sig=uJuWcb_aCuVqOZX5QzusS4VXAJs&hl=en&ei=wsumSoufO8rBtwey4JStCA&sa=X&oi=book_result&ct=result&resnum=1#v=onepage&q=Great%20Seals%20of%20Georgia%3B%20Their%20Origin%20and%20History&f=false>

Greene, Jack P. "Trevails of an Infant Colony: The Search for Viability, Coherence, and Identity in Colonial Georgia." *Imperatives, Behaviors & Identities: Essays in Early American Cultural History*. University Press of Virginia, Charlottesville, Va.: 1992. 113–141.

Hakylut, Richard. *Divers Voyages Touching the Discovery of America and the Islands Adjacent* (1582), ed. John Winter Jones. The Hakluyt Society, London: 1850. Reprint Elibron Classics, Boston: 2005.

Hamilton, Alexander, John Jay and James Madison. *The Federalist Papers*. Digital text Chris Whitten. <http://www.foundingfathers.info/federalistpapers/>

Hemperley, Marion. *The Georgia Surveyor General Department: A History and Inventory of Georgia's Land Office*. State Printing Office, Atlanta: 1982.

Hemperley, Marion. *Historic Indian Trails of Georgia*. Garden Club of Georgia, Athens, Ga.: 1989.

Hemperley, Marion R. and Edwin L. Jackson. *Georgia's Boundaries: The Shaping of a State*. University of Georgia Press, Athens, Ga.: 1972. (ISBN 978-08985-4142-7)

"Hernando de Soto." *Microsoft Encarta Online Encyclopedia*: 2007. <http://encarta.msn.com/encyclopedia_761574754/Hernando_de_Soto.html>

Hofstra, Warren. *The Planting of New Virginia: Settlement and Landscape in the Shenandoah*. John Hopkins University Press, Baltimore: 2005. (ISBN 080-1874-181)

Holmes, Jack D.L. "The Southern Boundary Commission: The Chattahoochee River and the Florida Seminoles, 1799," *Florida Historical Quarterly*, vol. 44, no. 4 (Apr. 1966): p. 312–341. <http://fulltext10.fcla.edu/DLData/SN/SN00154113/0044_004/44no4.pdf>

Hood, Jack Brian. "Georgia's Northern Boundary," *Georgia State Bar Journal*, vol. 8, no. 197 (1971).

Howard v. Ingersoll. 54 U.S. 381, 1851. Digital text Open Jurist. <http://openjurist.org/54/us/381/john-howard-peaintiff-in-error-v-stephen-m-ingersoll#fn-s>

Hvidt, Kristian, ed. *Von Reck's Voyage*. Beehive Press, Savannah, Ga.: 1980. (ISBN 0-88322-002-4)

Jackson, Harvey H. and Phinizy Spalding, eds. *Forty Years of Diversity: Essays on Colonial Georgia.* University of Georgia Press, Athens, Ga.: 1984. (ISBN 0-8203-0705-x)

Kluger, Richard. *Seizing Destiny: How America Grew from Sea to Shining Sea.* Alfred A. Knopf Publishing Co., New York: 2007. (ISBN 978-03754-1341-4)

Krakow, Kenneth K. *Georgia Place Names*, 3d ed. Winship Press, Macon, Ga.: 1999. (ISBN 0-915430-00-2)

Lannen, Andrew C. "Liberty and Authority in Colonial Georgia, 1717–1776." Ph.D. dissertation. Louisiana State University, Dec. 2002.

Linklater, Andro. *The Fabric of America: How Our Borders and Boundaries Shaped the Country and Forged Our National Identity.* Walker & Co., New York: 2007. (ISBN 100-8027-1533-8)

The Magazine of History with Notes and Queries, Vol. XIV. William Abbatt, New York: July-Dec. 1911.

Mathews, Catharine Van Cortlandt. *Andrew Ellicott: His Life and Letters.* Grafton Press, New York: 1908. Reprint World-Comm, Alexander, N.C.: 2001. (ISBN 1-56664-188-8)

Mitchell, John. "A map of the British and French dominions in North America, with the roads, distances, limits, and extent of the settlements, humbly inscribed to the Right Honourable the Earl of Halifax, and the other Right Honourable the Lords Commissioners for Trade & Plantations, by their Lordships most obliged and very humble servant, Jno. Mitchell. Tho: Kitchin, sculp." Hand colored, 2d ed. [London]: Sold by And: Mill[ar], 1755 [i.e. 1757]. Digital reproduction Library of Congress, Geography and Map Division, Washington, D.C. <http://hdl.loc.gov/loc.gmd/g3300.ar004000>

Mueller, Edward A. *Perilous Journeys: A History of Steamboating on the Chattahoochee, Apalachicola and Flint Rivers, 1828–1928.* Historic Chattahoochee Commission, Eufaula, Al.: 1990. (ISBN 0-945477-09-0)

North Carolina, State of. *North Carolina General Statutes*, §141-1: 2008. Digital text North Carolina General Assembly. <http://www.ncleg.net/EnactedLegislation/Statutes/HTML/ByChapter/Chapter_141.html>

North Carolina, State of. *Revised Statutes of the State of North Carolina passed by the General Assembly at the Session of 1836–7.*

Padgett, James A., ed. "Commission, Orders and Instruction Issued to George Johnstone, British Governor of West Florida, 1763–1767," *The Louisiana Historical Quarterly*, vol. 21, no. 4 (Oct. 1938): p. 1021–68.

Paine, Thomas. *The Crisis*. Digital text Independence Hall Asssociation, Philadelphia. <http://www.ushistory.org/Paine/crisis>

Paine, Thomas. *Common Sense, the Rights of Man and Other Essential Writings*. W. & T. Bradford, Philadelphia: 1776. Digital text Bartleby.com, New York: 1999. (ISBN 1-58734-037-2) <http://www.bartleby.com/133/>

Peabody, William and George Ellis. *Makers of American History: William Penn and James Oglethorpe*. The University Society, New York: 1904.

The Pinckney Treaty, 1795. Digital text "Treaty of Friendship, Limits, and Navigation between Spain and The United States; October 27, 1795," *Avalon Project*, 2008. Lillian Goldman Law Library, Yale Law School, New Haven, Conn. <http://avalon.law.yale.edu/18th_century/sp1795.asp>

"Records of the Settlers at the Head of the French Broad River, 1793–1803," *American Historical Review*, vol. 16, no. 4 (July 1911): p. 791–793. <http://www.journals.uchicago.edu/action/jstor?doi=10.2307%2F1835709>

Royal Proclamation, 1763. Digital text "The Royal Proclamation, 1763," *The Solon Law Archive*, 1994–2004. William F. Maton. <http://www.solon.org/Constitutions/Canada/English/PreConfederation/rp_1763.html>

Royal Proclamation, 1763. Digital text "The Royal Proclamation — October 7, 1763," *Avalon Project*, 2008. Lillian Goldman Law Library, Yale Law School, New Haven, Conn. <http://avalon.law.yale.edu/18th_century/proc1763.asp>

Oglethorpe, James. "A New and Accurate account of the Provinces of South Carolina and Georgia," 1732. In *The Most Delightful Country of the Universe: Promotional Literature of the Colony of Georgia, 1717–1734*, ed. Trevor R. Reese, p. 120–121. Beehive Press, Savannah, Ga.: 1972.

Sanders, Brad. *Guide to William Bartram's Travels*. Fevertree Press, Athens, Ga.: 2002. (ISBN 0-9718763-0-4)

"Savannah, from a print of 1741." Illustration in *A Popular History of the United States*, William Cullen Bryant and Sydney Howard Gay. Charles Scribners' Sons, New York: 1881. Digital reproduction U.S. History Images, 2009. <http://ushistoryimages.com/colonial-georgia.shtm>

Saye, Albert B. ed. "Commission and Instructions of Governor John Reynolds, August 6, 1754," *The Georgia Historical Quarterly*, vol. 30 (June 1946).

Sherwood, Adiel. *A Gazetteer of the State of Georgia*, 3d ed. P. Force, Washington D.C.: 1837. Reprint Genealogical Publishing Co., Baltimore: 2002.

Skaggs, Marvin L. *North Carolina Boundary Disputes Involving Her Southern Line*. University of North Carolina Press, Chapel Hill, N.C.: 1941.

Smith, George Gillman. *Story of Georgia and the Georgia People, 1732–1860*. Clearfield Co., Atlanta: 1900. (ISBN 0-8063-0317-4)

South Carolina, State of. *South Carolina Code of Laws (Unannotated)* §1-1-10: 2008. Digital text South Carolina Legislature. <http://www.scstatehouse.gov/code/t01c001.htm>

South Carolina v. Georgia. 93 U.S. 4 23, 1876. Digital text Justia & Oyez & Forms WorkFlow. <http://supreme.justia.com/us/93/4/case.html>

South Carolina v. Georgia. 93 U.S. 4, 1876. Digital text Open Jurist. <http://www.openjurist.org/93/us/4>

Spies, Gregory. "Major Ellicott's Triangulation: Establishing the Boundary between the Choctaw and the Creek Nations across the Mobile–Tensaw River Delta, 1798," History Corner, *Professional Surveyor Magazine*, vol. 24, no. 8 (Aug. 2004). <http://www.profsurv.com/magazine/article.aspx?i=1288>

Spies, Gregory. "Major Ellicott's Triangulation," History Corner, *Professional Surveyor Magazine*, vol. 24, no. 9 (Sept. 2004). <http://www.profsurv.com/magazine/article.aspx?i=1303>

Spies, Gregory. "The Mystery of the Camak Stone," History Corner, *Professional Surveyor Magazine*, vol. 24, no. 3 (Mar. 2004). <http://www.profsurv.com/magazine/article.aspx?i=1215>

Stocks, Thomas. "Memorandum made during my Tour in running the Dividing Line between Georgia & Tennessee, Commencing 5, May 1818." In "The Georgia-Tennessee Boundary Line," E. Merton Coulter, and *North Carolina Boundary Disputes Involving Her Southern Line*, Marvin L. Skaggs.

Stuart, Charles B. D. *Lives and Works: Civil and Military Engineers of America*. Van Nostrand, New York: 1871.

Sullivan, Buddy. *Georgia: A State History*. Arcadia Publishing, Mt. Pleasant, S.C.: 2003. (ISBN 0-7385-2408-5)

Tennessee, State of. House Bill 749: Apr. 8, 1889. In "The Georgia-Tennessee Boundary Line," E. Merton Coulter, and *North Carolina Boundary Disputes Involving Her Southern Line*, Marvin L. Skaggs.

Tennessee, State of. House Joint Resolution 919: 2008. Digital text Tennessee Department of State. <http://state.tn.us/sos/acts/105/resolutions/HJR0919.pdf>

Tennessee, State of. *Tennessee Code*: Oct. 30, 1819. In "The Georgia-Tennessee Boundary Line," E. Merton Coulter, and *North Carolina Boundary Disputes Involving Her Southern Line*, Marvin L. Skaggs.

Tennessee, State of. *Tennessee Code Annotated*, §4-2-105: 2008. Digital text Michie's Legal Resources. <http://www.michie.com/tennessee/lpext.dll?f=templates&fn=main-h.htm&cp=tncode>

Thomas, Keith Vivian. "James Edward Oglethorpe, 1696-1785." Chapel, Corpus Christi College, Oxford, England: 5 Oct. 1996.

"Trade, Board of." *Encyclopaedia Britannica Online*, 2008. <http://www.britannica.com/ed/article-9073132>

Treaty of Paris, 1763. Digital text "Treaty of Paris 1763," *Avalon Project*, 2008. Lillian Goldman Law Library, Yale Law School, New Haven, Conn. <http://avalon.law.yale.edu/18th_century/paris763.asp>

Treaty of Paris, 1783. Digital text "British-American Diplomacy: The Paris Peace Treaty of September 30, 1783," *Avalon Project*, 2008. Lillian Goldman Law Library, Yale Law School, New Haven, Conn. <http://avalon.law.yale.edu/18th_century/paris.asp>

Tubbs, Stephanie Ambrose and Clay Straus Jenkinson. *The Lewis and Clark Companion*. Henry Holt and Co., New York: 2003. (ISBN 978-08050-6726-2)

United States. National Register of Historic Places, "Ellicott Rock." Nomination form: July 24, 1973. Digital reproduction *National Register Sites in South Carolina*. <http://www.nationalregister.sc.gov/oconee/S10817737004/S10817737004.pdf>

United States of America v. 450 Acres of Land, More or Less, Known as Barnwell Island. 220 F.2d 353, cert. denied; 350 U.S. 826,1955. Digital text Open Jurist. <http://openjurist.org/220/f2d/353>

Vanderhill, B.G. and F.A. Unger. "The Georgia–Florida Land Boundary: Product of Controversy and Compromise," *The Southeastern United States: Essays on the Cultural and Historical Landscape*, vol. 18. West Georgia College, Carrollton, Ga.: 1979.

Whitaker, A.P. "The Muscle Shoals Speculation, 1783-1789," *The Mississippi Valley Historical Review*, vol. 13, no. 3 (Dec. 1926).

Whitaker, A.P. "Spanish Intrigue in the Old Southwest: An Episode, 1788-89," *The Mississippi Valley Historical Review*, vol. 12, no. 2 (Sept. 1925).

Index

Pages in *italics* indicate illustrative material.

A

Acadians, 31
acquiescence, doctrine of, 124–126, 142–143, 144
Adams, John, 45, 47, 54, 62–63, 75
Adams, John Quincy, 89
Adams, Samuel, 44
Adams–Onis Treaty (1819), 90, 100
adverse possession, doctrine of, 124
Agreement and Cession of Lands, Articles of (1802), 79, *79*, 84, 99, 103, 114, 115
Aiken, William, 26
Aix-la-Chapelle, Treaty of, 30
Alabama
  Alabama Territory, 84
  boundary disputes with Georgia, 111–116
  defining Georgia and, 84–88
Alabama v. Georgia (1859), 115–116
Albemarle Sound, 13
Allegheny Mountains, 7, 18, 21, 67, 80
Altamaha River, 20–21, *21*, 23, 33–34
The American Crisis (Paine), 49
American Indians. *See* Native Americans
American Legation Museum, 53
American Revolution, 47, 54–55, 58
Anne (frigate), 24
Apalachicola River, 39, 40, 41, 57, 74, 78
Appleton's Annual Cyclopedia for 1898, 120
Argall, Samuel, 12
Army Corps of Engineers, U.S., 138–140, 147
Articles of Agreement and Cession (1802), 79, *79*, 84, 94, 99, 114, 115
Athens (Georgia), 88–89
Atlanta, 41, 68–69, *69*
Augsburg, League of, 14
Augusta (Georgia), 26, 37, 49, 50, 52, 53, 59, 85, 89
Austrian Succession, War of, 30

B

Back River, 132–133, 135, 139, *147*
Back River Sediment Basin, 136
Bahamas, 3, 48
Baltimore, 11
Barnwell Island(s), 132, 142–144, 146
Battle of Bloody Marsh, 29, 152
Beaufort Convention (Treaty of Beaufort)
  birth of state and nation and, 60–62, *61*
  "nonexistent land" and, 80
  South Carolina/Georgia boundary dispute and, 135, 140, 141, 143, 145–146
Beck County (Tennessee), 121
birth date of Georgia, 63–64
Blackspear, General, 104
Bloody Marsh, Battle of, 29, 152
Blue Ridge, 129, 130–131
Board of Trade (England), 18–19, 21, 33–34, 41, 139
Board of Trade and Plantations (England), 55
Boston Marathon, 47
Boston Tea Party, 45
Boswell, James, 29
boundaries of United States, 57–58
Bourbon County (Georgia), 59–60
Branch Bank of the State of Georgia, 89
Brazil, Portuguese settlement of, 5
Brevard (North Carolina), 81
Brunswick (Georgia), 13, 37
Buchanan, John P., 119
Buncombe County (North Carolina), 128–129

C

Cabot, John, 5, 7, 9

Cabral, Pedro, 5
calendars, differences in, 63–64
Calloway, Colin, 31
Camak House, 89
Camak, James, 84, 85–86, 87–89, 116, 117, 118, 152
Canada, exploration of, 7–8
Canton (Georgia), 22–23
Cape Cod, 11
Caribbean, 5–6, 28, 30, 48
Carolana land grant, 12–13
Carolina, early settlement of, 12–14. *See also* North Carolina; South Carolina
Carter, Jimmy, 122
Cartier, Jacques, 7–8
Cashiers (North Carolina), 61, 80
Cathedral of San Juan Bautista, 6
Champlain, Samuel de, 8
Channel Chart (Examination Survey charts, Savannah Harbor Deepening Project), 132–134, 136, 138
Charles I (king of England), 12–13
Charles II (king of England), 13
Charles Towne, 13–14. *See also* Charleston
Charles V (king of Spain), 6
Charleston, 17, 18, 19, 24, 27, 29, 50, 53. *See also* Charles Towne
Chatham, Earl of (William Pitt), 52
Chattahoochee River
  Alabama/Georgia boundary and, 111–113, 115–116
  Florida/Georgia boundary and, 40–41, 98, 99–105
  Georgia land cession and, 80
  legal boundaries of Georgia and, 95–97, *95*
  Mississippi, 77–78
  U.S. border, 57, 74
Chattanooga (Tennessee), 88, 120, 125
Chattooga River, 61, *61*, *82*, 82–83, 86, *87*, 94–96, *95*, 116, 126, 131, 135, 141–142
Cherokee people, 1, 84, 90, 118
Chesapeake Bay, 7, 54
Chickasaw people, 77
Chinese settlers in Georgia, 22–23
Churchill, Winston, 30
Coastal Zone Management Act of 1972 (U.S.), 145
Cocke, John, 84
Code of 1819 (Tennessee), 86

Coercive Acts (Intolerable Acts), 45
Coffee v. Groover (1887), 101–111
Columbus, Christopher, 3–4, 5–6, 69
Commerce Clause of U.S. Constitution, 140
commercial fishing, 145, 148
Commissioner's Rock, 83
Committee of Correspondence, 44
Common Sense (Paine), 49
Concord, battle of, 47
"Concord Hymn" (Emerson), 47
condemnation proceedings, 142–144
Confederation Congress, 62, 67, 93
Confederation Powder Works, 85
Congress, U.S., 52, 60, 84
constitution of Georgia, 52, 77–78, 94
Constitution, U.S., 52, 62, 98, 140
Constitutional Convention, 52
Continental Army, 48
Continental Association, 46
Continental Congress
  First Continental Congress, 45
  Georgia's constitution and, 52
  renaming of, 62
  Second Continental Congress, 46, 48–49
  struggle for independence and, 54
"continentals" (paper money), 48
Coram, Thomas, 18
Cornwallis, Charles, 54
Council of Safety, 48–49, 50, 51, 53
Creek people, 24–25, 34, 35, 84, 89, 102, 103, 118
"Croatoan," 10
Cuba, 6, 28, 32
Cumberland Island, 25, 75

D

Dade County (Georgia), 118, 120–121, 122
Dare, Virginia, 10
Darien (Georgia), 26, 37
Daytona Beach (Florida), 13
de jure and *de facto*, 96–97
De La Warr, Lord, 11–12
de León Fandiño, Julio, 28
de Soto, Hernando, 6–7
"debatable lands," 20, 28, 32, 34
debtors in Georgia, 24
"Declaration and Resolves of the First Continental Congress," 45–46

Declaration of Independence, 49–51
Declaratory Act of 1766 (England), 44
Denwill shoal, 146–147
Divers Discovery Touching America and the Islands Adjacent (Hakluyt), 9
drought of 2008, 1, 122–123, 125

E

early exploration, 3–16
  Carolina and, 12–14
  early explorers from Spain and Portugal, 3–7
  England and, 5, 8–10
  France and, 7–8
  important dates, 16
  Intercolonial Wars, 14–15
  Jamestown and, 11–12
  Native Americans and, 15
  Plymouth Company and, 11
  Virginia and, 10
"earmarks," 52
East Florida, colony of, 40, *40*, 45, 48, 100, 102, 106. *See also* Florida
East India Trading Company, 9, 45
Ebenezer (Georgia), 26–27
Echota, Treaty of (1835), 90
Elizabeth I (queen of England), 9–10
Ellicott, Andrew Jr., 72, 74
Ellicott, Andrew Sr.
  compared with Camak, 88
  Jefferson and, 70, 73, 75, 152
  "nonexistent land" and, 82–84
  southern boundary of Georgia and, 72–76, 100, 102, 104, 106, 109
  Washington and, 72, 152
Ellicott Stone, 73
Ellicott's Mound, 75, 89–90 90, *95*, 96, 98–101, 102–105, 108–111, *110*
  McNeil Line and, 100–101
Ellicott's Rock, *82*, 82–83, *87*, 88, 130
Ellis, Henry, 34
Emerson, Ralph Waldo, 47
England
  early exploration by, 5, 8–10
  Intercolonial Wars and, 30–31
Eton (college, England), 9, 19
Examination Survey charts, Savannah Harbor Deepening Project, 132–134, 136, 138
exploration. *See* early exploration

F

Fannin County (Georgia), 121
Federal Territory, 70
"The Federalist Papers" (Hamilton), 62
First Continental Congress, 45
First Provincial Congress. *See* Provincial Congresses of Georgia
fishing, commercial, 145, 148
Flint River, 40–41, 74, 94–97, *95*, 98–100, 102–103, 105, 108, 111–112
Florida
  boundary disputes with Georgia, 98–111
  cession of Spanish Florida, 89–90
  defining Georgia and, 89–90
  East Florida, colony of, 40, *40*, 45, 48, 100, 102, 106
  naming of, 6
  Proclamation Line and, 38–42
  West Florida, colony of, 39, *39*, 42, *42*, 45, 48, *71*, 100, 102, 106
Florida Gravel Co. v. Capital City Sand & Gravel Co. (1930), 99–100
Floyd, General, 104
Fort Frederica (Georgia), 28, 29, 74
Fort St. Simon (Georgia), 29
Foundling Hospital (London), 18
450 Acres of Land, More or Less, Known as Barnwell Island, United States v. (1955), 142–144
France
  early exploration by, 7–8
  Intercolonial Wars and, 30–31
Francis I (king of France), 7–8
Franco–American Alliance, 53
Franklin, Benjamin, 54, 152
Frazier, James, 120–121
French and Indian War, 30–31, 53, 58

G

Gaines, James, 84, 85–86, 116
General Assembly of Georgia, 59, 83
Genius of the Colony figure, 23
Geological Survey, U.S., 143, 144
George II (king of England)
  Augusta's naming and, 26
  death of, 38
  Georgia named after, 18, 19
  sponsorship of 13th colony, 17
  Trustees of Georgia and, 1, 3, 33, 63

177

Yamacraw tribe and, 25
George III (king of England), 38, 45–46, 48, 54, 101–102
Georgia: Comprising Sketches of Counties, Towns, Events, Institutions and Persons in Cyclopedic Form, 120
Georgia Constitution, 52, 77–78, 94
"Georgia Day," 63–64
Georgia/North Carolina boundary line commission, 122
Georgia Railroad, 89
"Georgia/South Carolina Boundary Project, Lower Savannah River Segment, Portfolio of Maps," 135
Georgia/Tennessee boundary line commission, 122, 124, 125
Georgia v. South Carolina (1922), 124, 140–142
Georgia v. South Carolina (1990), 1, *146*, 146–148
Georgia's Boundaries: The Shaping of a State (Hemperley & Jackson), 151
Gilbert, Humphrey, 9
Gilbert, John, 9
Gilbert, Raleigh, 9–10
Grand Alliance, War of the, 14
Great Locomotive Chase, 118
Gregorian calendar, 63
Gregory XIII (pope), 63
Grenada, colony of, 38
Gulf of Mexico, 6, 30, 39, 40, 41, 73, 74, 76
Gunter, Edmund, 68
Gunter's chain, 68
Gwinnett, Button, 49, 50–51, 60

H

Hakluyt, Richard, 9, 10, 152
Hall, Lyman, 46–47, 49–50
Hamilton, Alexander, 62
Hamilton County (Tennessee), 118, 120–121, 122
Hancock, John, 44, 45, 47
"The Hard Case of the Distressed People of Georgia" (Stephens, Thomas), 27
Harrison, John, 69–70
Hartsfield-Jackson Atlanta International Airport, 40–41
Heather, Robert, 12–13
Hemperley, Marion, 151
Henry, Patrick, 12, 45

Henry VII (king of England), 5
high water mark and low water mark, importance of, 115–16
Hilton Head Island, 24, 134, 138, 145, 147, *147*
Hispaniola, 4, 6
Hoffman, Walter E., 146, 147
Holland, Georgia's independence and, 53–54
Horseshoe shoal, 146–47, *147*
House Joint Resolution 919, Tennessee ("A resolution relative to the Tennessee–Georgia boundary"), 124–125
House of Burgesses (Virginia), 12
Houston County (Georgia), 59
Howard, John H., 112–113
Howard v. Ingersoll (1851), 112–115
Hutchinson Island, 132, 135–136

I

independence for Georgia, 45–49, 53–55
Indians, 4. *See also* Native Americans
Ingersoll, Stephen M., 112–13
Intercolonial Wars, 14–15, 30–32, 43
International Meridian Conference, 70–71
Intolerable Acts (Coercive Acts), 45
Iroquois people, 7, 14
Isabella (queen of Spain), 3, 5
Italy, silk production in, 22

J

Jackson, Andrew, 89–90, 105
Jackson, Ed, 151
Jackson, James, 50
James I (king of England), 11, 12
Jamestown, 11–12
Jay, John, 45, 54, 62
Jefferson Stone, 70
Jefferson, Thomas
  Declaration of Independence and, 49
  Ellicott and, 70, 73, 75, 152
  in House of Burgesses, 12
  land expansion and, 67
Jenkins' Ear, War of, 28–30
Jenkins, Robert, 28
Johnson, Robert, 18, 24
Johnson, Samuel, 29
Johnstone, George, 42
Jones Island, 133–134, 137–138, 147

The Journal of Andrew Ellicott, for Determining the Boundary between the United States and the Possessions of His Catholic Majesty in America (Ellicott), 76, 83–84
Journal of the Transactions of the Trustees, first mention of name Georgia in, 63
Julian calendar, 63
Julius Caesar, 63
jurisdiction, original, 98, 120

K

Keith, William, 18–19
Keowee River, *58*, 60–61, *61*, 77, 94, 140–141
King George's War, 30
King William's War, 14

L

laches, doctrine of, 125
Lachine Rapids (Canada), 8
Lake Randolph, 100, 106
Lake Seminole, 97, 100
Land Acts of 1784, 1785 and 1787 (U.S.), 67
land cession, 67, 79–80. *See also* Articles of Agreement and Cession
land lotteries, 117–118
lateral seaward boundary, 145–146, 147, *147*, 148
latitude and longitude, 68–71, *69*
League of Augsburg, 14
Lexington, battle of, 47
Line of Demarcation, 4–5
Longfellow, Henry Wadsworth, 47
longitude. *See* latitude and longitude
Lookout Mountain, 120
Lords Proprietors, 13–14, 17–18
"The Lost Colony" (drama), 10
Louis XIV (king of France), 14
Louisiana, 31–32, 53, 58
Love, Robert, 86, 130
low water mark and high water mark, importance of, 115–116
Lutherans, 26–27

M

Madison, James, 62
Malcontents, 27–28, 152

"Map of the British and French Dominions in North America" (Mitchell, John), 55
Margravate of Azilia, 17–18
Marion County (Tennessee), 118, 120–121, 122
Mason Dixon line, 72
Massachusetts, in Revolutionary War, 47
Massachusetts Bay Colony, 11
Massachusetts Militia, 47
Matthews, George, 59
McBride, John, 106
McCord, Jim, 122
McIntosh, Lachlan, 51, 60
McMinn, Joseph, 84
McNeil, D. F., 100, 105
McNeil Line, 101, 105, 109
Memorandum made during my Tour in running the Dividing Line between Georgia & Tennessee, Commencing 5, May 1818 (Stocks), 85
Menéndez de Avilés, Pedro, 7
metes and bounds, 93
"The Midnight Ride of Paul Revere" (Longfellow), 47
Milledgeville (Georgia), 81, 83, 89
Miller's Bend, *79*, 95, 97, 111–112, 115
Minor, Stephen, 102, 104, 105, 106, 109
Mississippi, defining Georgia and, 77–78
Mississippi River, 6–7, 8, 15, 31–32, 39, 42, 55, 57, 58, 59, 67, 71–72, 72–73, 76, 77–78, 102–103, 141
Mississippi Territory, 77, 78–79
missions, Spanish, 7, 15, 25, 32
Mitchell, David B., 82, 84
Mitchell, John, 55
Mitchell Map, 55, *56*
Mobile (Alabama), 31, 73
Mobile River, 73, 75
Mohawk people, 14, 58
Montgomery, Hugh, 84, 87
Montgomery, Robert, 17–18
Montgomery's Corner, 84–88, *87*, *95*
Montgomery's Line, 131
Montreal, 8, 14, 31
Morocco, recognition of U.S. by, 53
mulberry trees, 22, 27
Musgrove, Mary, 24

N

National Oceanic and Atmospheric Administration (NOAA), 134, 135
National Register of Historic Places, 73, 83
Native Americans
  Cherokee, 1, 84, 90, 118
  Chickasaw, 77
  Creek, 24–25, 34, 35, 84, 89, 102, 103, 118
  de Soto and, 6–7
  early colonization and, 15
  Iroquois, 7, 14
  land cession treaties and, 67
  Mohawk, 14, 58
  Plymouth Company and, 11
  Proclamation Line and, 38
  trade with, 139
  Yamacraw, 24–25
Natchez (Mississippi), 59, 72–73, 102
Netherlands, Georgia's independence and, 53–54
New Orleans, 8, 31, *39*, 58, 71–72, 73, 89
New York City, 11, 62
New York Independent Journal, "The Federalist Papers" published in, 62
Newfoundland, 5, 7, 8, 9, 55
Nickajack (Native American village), 79, 84–88, *87*, 95, 97, 99, 111–112, 115, 116, 126
Nickajack Cave, 84–85
Nickajack Lake, 84, 97
"no taxation without representation," 44
NOAA (National Oceanic and Atmospheric Administration), 134, 135
"nonexistent land" dispute, 80–84, *81*
North Carolina
  boundary disputes with Georgia, 126–131
  establishment of, 14
  "nonexistent land" dispute and, 80–84
North, Lord, 54
Northern, William, 119
Northwest Passage, 7, 8, 9
Nova Scotia, 5, 7, 8, 14, 17, 18, 30–31, 57–58

O

Official Code of Georgia Annotated (O.C.G.A.)
  Alabama/Georgia boundary in, 111
  Florida/Alabama boundary in, 98–99
  legal boundaries of Georgia and, 94–96, 97
  North Carolina/Georgia boundary in, 126
  South Carolina/Georgia boundary in, 131–34
  Tennessee/Georgia boundary in, 116
Oglethorpe, James
  arrival in Georgia, 63
  biographical details of, 19
  commemoration of, 29–30
  dissent and Malcontents and, 27–28
  failure of Georgia colony and, 24, 152
  "Georgia Day" and, 64
  McIntosh and, 51
  meeting with Adams (John), 54
  new colony of Georgia and, 23–26
  War of Jenkins' Ear and, 28–30
oil, offshore, 145
Okefenokee swamp, 74–75, 89, 102–103
Oklahoma v. Texas (1922), 124
Olive Branch Petition, 48
original jurisdiction, 98, 120
Orr, Gustavus J., 96, 98, 108–109
Orr–Whitner Line, 108–111, *110*
Outer Banks (North Carolina), 13
Oyster Bed Island, 133, 137, 146, *147*

P

Paine, Thomas, 49, 152
Paris, Treaty of (1763), 31–32, *32*, 58, 97
Paris, Treaty of (1783), 55–56
Patriots' Day, 47
Peace and Friendship, Treaty of (Morocco–U.S.), 53
Pearl River, 73
Pennyworth Island, 132, 135–36
Pensacola (Florida), *39*
Perdue, Sonny, 123
Philadelphia
  Ellicott and, 72, 75, 82
  in Revolutionary War, 45, 48–49, 62
Pilgrims, 11
Pinckney, E. B., 142–143, 144
Pinckney, Thomas, 72
Pinckney Treaty, 71–72, 74, 89, 99
Pitt, William (Earl of Chatham), 52
Pizarro, Francisco, 6

Plymouth colony, 11
Plymouth Company, 11
Pocahontas, 12
Polk County (Tennessee), 121
Polo, Marco, 3
Ponce de León, Juan, 6
Popham Colony, 11
"pork barrel," 52
Port Royal (South Carolina), 24
Portugal
  early explorers from, 3–7
  settlement of Brazil, 5
prescription, doctrine of, 125–26, 142–143
prime meridians, 70–71
Proclamation Act of 1763 (England), 38
Proclamation Line, *38*, 38–42
Provincial Congresses of Georgia, 46, 48, 49, 50, 52
Purry, Jean Pierre, 18
Purrysburg (South Carolina), 18

Q

Quartering Act of 1765 (England), 43–44
Quebec, colony of, 38, 45, 48
Quebec City, 8, 31
Queen Anne's War, 15

R

Raleigh, Walter, 10
Randolph, Thomas M., 105
Rebecca (brig), 28
Revere, Paul, 48
Revised Statutes of The State of North Carolina passed by the General Assembly at the Session of 1836-37, 127–131
Revolutionary War, 47, 54–55, 58
Reynolds, John, 33–34
Reynolds, Joshua, 29
Rhode Island, 17, 62
Ribault, Jean, 41
riparian rights, 93, 114
Roanoke colony, 10
Robertson, William, 9
Rolfe, John, 12
Royal General Assembly, 46
Royal Greenwich Observatory, 71
Royal Navy, 54
Royal Period of Georgia, 32–35
Royal Regulars, 47, 54

rum, Georgia's dealings with, 21, 27, 28, 33, 139
Rye, Tom C., 121
Ryswick, Treaty of, 14

S

St. Augustine, 7, 13, 20, 28, *40*, 53
St. John's Island, 9
St. John's Parish, 46, 50, 52
Saint Lawrence River, 7–8, 57
St. Marks (Florida), 74
St. Marys (village, Florida), 74–75
St. Marys River, 34, 39, 40, 41–42, 74–75, *95*, 96, 98–106, *110*
St. Simons Island, 28, 29, 74
Sally (surveying boat), 73, 74, 75
Salzburgers, 26–27
Savannah (Georgia), 25, 29–30, 37, 46, 48, 49, 50, 51, 52, 53, 55, 63, 75, 82, 139–140, 142, 145, *147*. See also Savannah Towne
Savannah Harbor, 142
Savannah Harbor Deepening Project, Examination Survey charts for, 132–134, 136, 138
Savannah River
  new colony of Georgia and, 20, *21*
  in Seal of Georgia, 23
  South Carolina/Georgia boundary and, 60–61, 94–96, *95*, 131–34, 135–38, 138–142, 143, 147–148, *147*
Savannah Towne (Georgia), 25–26, *26*. See also Savannah
Scotland, 51, 55
Scottish immigrants, 26
The Scratch of a Pen: 1763 and the Transformation of North America (Calloway), 31
Seal of Georgia, 23, *23*
Second Continental Congress, 46, 48–49
Second Provincial Congress. *See* Provincial Congresses of Georgia
Seven Years' War, 30
silk production, 22–23, 27
"single subject rule," 52
Slaton, John M., 121
smugglers and smuggling, 44
Sons of Liberty, 44–45
South Carolina
  Beaufort Convention and, 60–62
  boundary disputes with Georgia, 131–

148
  establishment of, 14
  1977 resolution in boundary dispute with Georgia, 144
South Carolina v. Georgia (1876), 139–140
Southern Cultivator (journal), Camak as editor of, 89
Spain
  early explorers from, 3–7
  Georgia's independence and, 53, 54
  Jackson's attack on Spanish Florida, 89–90
  Pinckney Treaty and, 71–72
  Treaty of Paris (1763) and, 31–32
  War of Jenkins' Ear and, 28–29
Spanish missions, 7, 15, 25, 32
Spanish Succession, War of, 14–15
Spaulding, Thomas, 106
spoil islands, 142, 146, 148
Stamp Act of 1765 (England), 43, 44
"A State of the Province of Georgia" (Stephens, William), 27
Stephens, Thomas, 27–28, 152
Stephens, William, 27, 28
Stocks, Thomas, 84, 85
Sugar Act of 1764 (England), 43, 44
Sugar and Molasses Act (England), 43
Supreme Court (U.S.)
  Alabama/Georgia boundary disputes in, 112–116
  Florida/Georgia boundary disputes in, 100–111
  original jurisdiction of, 98
  South Carolina/Georgia boundary disputes in, 138–42, 145–146
  Yazoo Land Fraud and, 76
surveying, 68, 72–76, 93

T

Tallulah River, 61
Talmadge, Eugene, 122
taxation in American colonies, 43–44
Tea Act of 1773 (England), 45
Tennessee
  boundary disputes with Georgia, 116–126
  defining Georgia and, 76–77, 84–88
  Georgia/Tennessee boundary line commission, 122, 124, 125
Tennessee River, 79, 80, 84–86, 87, 118,
123, 124
Third Provincial Congress. *See* Provincial Congresses of Georgia
13th colony, Georgia as, 17–36
  dissent and Malcontents, 27–28
  first settlements, 25–27
  five early attempts at, 17–19
  important dates, 36
  Intercolonial Wars and, 30–32
  as new colony, 20–24
  Oglethorpe and, 28–30
  relations with Creek, 24–25
  Royal Period of Georgia, 32–35
  Trustees of Georgia, 19–20
  War of Jenkins' Ear, 28–30
35th parallel, 77, *81*, 81–82, *82*, 84–86, *87*, 88, *95*, 96, 99, 116–117, 123–125, 126, 127, 129, 130–131, 131, 135, 141
31st parallel, 39, *41*, 71–72, *71*, 73–74, 77, *95*, 97, 99–100, 102, 112
32° 28' north, 42, *42*, 71, *71*, 77–78
Thomas Coram Foundation for Children, 18
Thompson, General, 104
tobacco, 12, 19, 22, 43
Tomochichi, 24–25
Toonahowi, 25
Tordesillas, Treaty of (1494), *4*, 4–5
Townshend Acts of 1767 (England), 44
Trail of Tears, 90
Treaty of Aix-la-Chapelle, 30
Treaty of Beaufort. *See* Beaufort Convention
Treaty of Echota (1835), 90
Treaty of Paris (1763), 31–32, *32*, 58, 97
Treaty of Paris (1783), 55–56
Treaty of Peace and Friendship (Morocco–U.S.), 53
Treaty of Ryswick, 14
Treaty of Tordesillas (1494), *4*, 4–5
Treaty of Utrecht, 15
Trinidad, 12
"A True and Historical Narrative of the Colony of Georgia" (paper by member of Malcontents), 27
Trustees of Georgia
  George II and, 1, 3, 33, 63
  Georgia as 13th colony and, 19–20, *21*, 21–22, 23–28
  South Carolina/Georgia boundary dispute and, 138

surrendering of colony by, 33
truth, stranger than fiction, 1–154
Tugaloo River, *58*, 60–62, 77–78, *78*, 94–95, 131, 135, 140–142
Turney, Peter, 119
Tybee Island, 49, 134, 138, 147, *147*
Tyrral, Timothy, 130

U

Uchee creek, 80, 112
Unicoi Mountain, 117
U.S. Board on Geographic Names, 34
United States Coastal Zone Management Act of 1972, 145
U.S. Geological Survey, 143–144
U.S. Postal Service, 30
United States v. 450 Acres of Land, More or Less, Known as Barnwell Island (1955), 142–144
University of Georgia, 50, 88–89
Utrecht, Treaty of, 15

V

Vicksburg (Mississippi), 42, *42*
Virginia and early exploration, 10
Virginia Company, 11, 12

W

Walton County (Georgia), 81, 128, 129
Walton, George, 49, 50, 81
War of Austrian Succession, 30
War of the Grand Alliance, 14
War of Jenkins' Ear, 28–30
War of Spanish Succession, 14–15
Washington, George
  Continental Army and, 48
  defeat of Royal Navy, 54
  Ellicott and, 72, 152
  First Continental Congress and, 45
  as first U.S. president, 63
  in House of Burgesses, 12
  land expansion and, 67
Watson, J. C., 104
Watson Line, 104–5, 110–111
West Florida, colony of, 39, *39*, 42, *42*, 45, 48, 71, 100, 102, 106. *See also* Florida
West Indies, 4, 13, 28, 38, 43
West Point (Georgia), 87, *95*, 97

Western and Atlantic Railroad, 118, 120
western expansion, 58–60
Westminster Abbey, 49
White, John, 10
Whitner, B. F., 96, 98, 108, 109
William, Duke of Cumberland, 25
Wilmington (North Carolina), 11
Wimbish Report (1898), 120
Wimbish, W. A., 120
Wormslow Plantation (Georgia), 29
Wright, Elizabeth, 29
Wright, James
  arguing for invasion of Georgia, 53
  Board of Trade and, 41, *41*
  Council of Safety and, 48
  First Provincial Congress and, 46
  as governor of Georgia, 34–35
  leaving Georgia, 49, 55
  legacy of, 152
Wrightsboro (Georgia), 49

Y

Yamacraw people, 24–25
Yazoo Land Fraud, 76, 79
Yorktown, battle of, 54

LaVergne, TN USA
14 April 2010
179262LV00001B/187/P